Susan Scherffius Jakes
Craig C. Brookins
Editors

Understanding Ecological Programming: Merging Theory, Research, and Practice

Understanding Ecological Programming: Merging Theory, Research, and Practice has been co-published simultaneously as *Journal of Prevention & Intervention in the Community*, Volume 27, Number 2 2004.

Pre-publication
REVIEWS,
COMMENTARIES,
EVALUATIONS . . .

"IMMENSELY USEFUL for advancing the theory and practical application of ecologically sound prevention programs. . . . Clarifies the often used but rarely defined adjective 'ecological' as applied to prevention programming. Researchers interested in designing and implementing ecological programs will find here a tool to test and insure that ecological principles are followed, as well as real-world examples of programs that vary in adherence to those principles."

Joel M. Hektner, PhD,
Assistant Professor
Child Development and Family Science,
North Dakota State University

Understanding Ecological Programming: Merging Theory, Research, and Practice

Understanding Ecological Programming: Merging Theory, Research, and Practice has been co-published simultaneously as *Journal of Prevention & Intervention in the Community*, Volume 27, Number 2 2004.

The *Journal of Prevention & Intervention in the Community*™ Monographic "Separates" (formerly the *Prevention in Human Services* series)*

For information on previous issues of *Prevention in Human Services*, edited by Robert E. Hess, please contact: The Haworth Press, Inc., 10 Alice Street, Binghamton, NY 13904-1580 USA.

Below is a list of "separates," which in serials librarianship means a special issue simultaneously published as a special journal issue or double-issue *and* as a "separate" hardbound monograph. (This is a format which we also call a "DocuSerial.")

"Separates" are published because specialized libraries or professionals may wish to purchase a specific thematic issue by itself in a format which can be separately cataloged and shelved, as opposed to purchasing the journal on an on-going basis. Faculty members may also more easily consider a "separate" for classroom adoption.

"Separates" are carefully classified separately with the major book jobbers so that the journal tie-in can be noted on new book order slips to avoid duplicate purchasing.

You may wish to visit Haworth's website at . . .

http://www.HaworthPress.com

. . . to search our online catalog for complete tables of contents of these separates and related publications.

You may also call 1-800-HAWORTH (outside US/Canada: 607-722-5857), or Fax 1-800-895-0582 (outside US/Canada: 607-771-0012), or e-mail at:

docdelivery@haworthpress.com

Understanding Ecological Programming: Merging Theory, Research, and Practice, edited by Susan Scherffius Jakes, PhD, and Craig C. Brookins, PhD (Vol. 27, No. 2, 2004). *Examines the background, concept, components, and benefits of using ecological programming in intervention/prevention program designs.*

Leadership and Organization for Community Prevention and Intervention in Venezuela, edited by Maritza Montero, PhD (Vol. 27, No. 1, 2004). *Shows how (and why) participatory communities come into being, what they can accomplish, and how to help their leaders develop the skills they need to be most effective.*

Empowerment and Participatory Evaluation of Community Interventions: Multiple Benefits, edited by Yolanda Suarez-Balcazar, PhD, and Gary W. Harper, PhD, MPH (Vol. 26, No. 2, 2003). *"USEFUL . . . . Draws together diverse chapters that uncover the how and why of empowerment and participatory evaluation while offering exemplary case studies showing the challenges and successes of this community value-based evaluation model." (Anne E. Brodsky, PhD, Associate Professor of Psychology, University of Maryland Baltimore County)*

Traumatic Stress and Its Aftermath: Cultural, Community, and Professional Contexts, edited by Sandra S. Lee, PhD (Vol. 26, No. 1, 2003). *Explores risk and protective factors for traumatic stress, emphasizing the impact of cumulative/multiple trauma in a variety of populations, including therapists themselves.*

Culture, Peers, and Delinquency, edited by Clifford O'Donnell, PhD (Vol. 25, No. 2, 2003). *"TIMELY . . . . OF VALUE TO BOTH STUDENTS AND PROFESSIONALS. . . . Demonstrates how peers can serve as a pathway to delinquency from a multiethnic perspective. The discussion of ethnic, racial, and gender differences challenges the field to reconsider assessment, treatment, and preventative approaches." (Donald Meichenbaum, PhD, Distinguished Professor Emeritus, University of Waterloo, Ontario, Canada; Research Director, The Melissa Institute for Violence Prevention and the Treatment of Victims of Violence, Miami, Florida)*

Prevention and Intervention Practice in Post-Apartheid South Africa, edited by Vijé Franchi, PhD, and Norman Duncan, PhD, consulting editor (Vol. 25, No.1, 2003). *"Highlights the way in which preventive and curative interventions serve–or do not serve–the ideals of equality, empowerment, and participation. . . . Revolutionizes our way of thinking about and teaching socio-pedagogical action in the context of exclusion." (Dr. Altay A. Manço, Scientific Director, Institute of Research, Training, and Action on Migrations, Belgium)*

Community Interventions to Create Change in Children, edited by Lorna H. London, PhD (Vol. 24, No. 2, 2002). *"ILLUSTRATES CREATIVE APPROACHES to prevention and intervention with at-risk youth. . . . Describes multiple methods to consider in the design, implementation, and evaluation of programs." (Susan D. McMahon, PhD, Assistant Professor, Department of Psychology, DePaul University)*

Preventing Youth Access to Tobacco, edited by Leonard A. Jason, PhD, and Steven B. Pokorny, PhD (Vol. 24, No. 1, 2002). *"Explores cutting-edge issues in youth access research methodology. . . . Provides a thorough review of the tobacco control literature and detailed analysis of the methodological issues presented by community interventions to increase the effectiveness of tobacco control. . . . Challenges widespread assumptions about the dynamics of youth access programs and the requirements for long-term success." (John A. Gardiner, PhD, LLB, Consultant to the 2000 Surgeon General's Report* Reducing Youth Access to Tobacco *and to the National Cancer Institute's evaluation of the ASSIST program)*

The Transition from Welfare to Work: Processes, Challenges, and Outcomes, edited by Sharon Telleen, PhD, and Judith V. Sayad (Vol. 23, No. 1/2, 2002). *A comprehensive examination of the welfare-to-work initiatives surrounding the major reform of United States welfare legislation in 1996.*

Prevention Issues for Women's Health in the New Millennium, edited by Wendee M. Wechsberg, PhD (Vol. 22, No. 2, 2001). *"Helpful to service providers as well as researchers . . . A USEFUL ANCILLARY TEXTBOOK for courses addressing women's health issues. Covers a wide range of health issues affecting women." (Sherry Deren, PhD, Director, Center for Drug Use and HIV Research, National Drug Research Institute, New York City)*

Workplace Safety: Individual Differences in Behavior, edited by Alice F. Stuhlmacher, PhD, and Douglas F. Cellar, PhD (Vol. 22, No. 1, 2001). Workplace Safety: Individual Differences in Behavior *examines safety behavior and outlines practical interventions to help increase safety awareness. Individual differences are relevant to a variety of settings, including the workplace, public spaces, and motor vehicles. This book takes a look at ways of defining and measuring safety as well as a variety of individual differences like gender, job knowledge, conscientiousness, self-efficacy, risk avoidance, and stress tolerance that are important in creating safety interventions and improving the selection and training of employees.* Workplace Safety *takes an incisive look at these issues with a unique focus on the way individual differences in people impact safety behavior in the real world.*

People with Disabilities: Empowerment and Community Action, edited by Christopher B. Keys, PhD, and Peter W. Dowrick, PhD (Vol. 21, No. 2, 2001). *"Timely and useful . . . provides valuable lessons and guidance for everyone involved in the disability movement. This book is a must-read for researchers and practitioners interested in disability rights issues!" (Karen M. Ward, EdD, Director, Center for Human Development; Associate Professor, University of Alaska, Anchorage)*

Family Systems/Family Therapy: Applications for Clinical Practice, edited by Joan D. Atwood, PhD (Vol. 21, No. 1, 2001). *Examines family therapy issues in the context of the larger systems of health, law, and education and suggests ways family therapists can effectively use an intersystems approach.*

HIV/AIDS Prevention: Current Issues in Community Practice, edited by Doreen D. Salina, PhD (Vol. 19, No. 1, 2000). *Helps researchers and psychologists explore specific methods of improving HIV/AIDS prevention research.*

Educating Students to Make-a-Difference: Community-Based Service Learning, edited by Joseph R. Ferrari, PhD, and Judith G. Chapman, PhD (Vol. 18, No. 1/2, 1999). *"There is something here for everyone interested in the social psychology of service-learning." (Frank Bernt, PhD, Associate Professor, St. Joseph's University)*

Program Implementation in Preventive Trials, edited by Joseph A. Durlak and Joseph R. Ferrari, PhD (Vol. 17, No. 2, 1998). *"Fills an important gap in preventive research. . . . Highlights an array of important questions related to implementation and demonstrates just how good community-based intervention programs can be when issues related to implementation are taken seriously." (Judy Primavera, PhD, Associate Professor of Psychology, Fairfield University, Fairfield, Connecticut)*

Preventing Drunk Driving, edited by Elsie R. Shore, PhD, and Joseph R. Ferrari, PhD (Vol. 17, No. 1, 1998). *"A must read for anyone interested in reducing the needless injuries and death caused by the drunk driver." (Terrance D. Schiavone, President, National Commission Against Drunk Driving, Washington, DC)*

Manhood Development in Urban African-American Communities, edited by Roderick J. Watts, PhD, and Robert J. Jagers (Vol. 16, No. 1/2, 1998). *"Watts and Jagers provide the much-needed foundational and baseline information and research that begins to philosophically and empirically validate the importance of understanding culture, oppression, and gender when working with males in urban African-American communities." (Paul Hill, Jr., MSW, LISW, ACSW, East End Neighborhood House, Cleveland, Ohio)*

Diversity Within the Homeless Population: Implications for Intervention, edited by Elizabeth M. Smith, PhD, and Joseph R. Ferrari, PhD (Vol. 15, No. 2, 1997). *"Examines why homelessness is increasing, as well as treatment options, case management techniques, and community intervention programs that can be used to prevent homelessness." (American Public Welfare Association)*

Education in Community Psychology: Models for Graduate and Undergraduate Programs, edited by Clifford R. O'Donnell, PhD, and Joseph R. Ferrari, PhD (Vol. 15, No. 1, 1997). *"An invaluable resource for students seeking graduate training in community psychology . . . [and will] also serve faculty who want to improve undergraduate teaching and graduate programs." (Marybeth Shinn, PhD, Professor of Psychology and Coordinator, Community Doctoral Program, New York University, New York, New York)*

Adolescent Health Care: Program Designs and Services, edited by John S. Wodarski, PhD, Marvin D. Feit, PhD, and Joseph R. Ferrari, PhD (Vol. 14, No. 1/2, 1997). *Devoted to helping practitioners address the problems of our adolescents through the use of preventive interventions based on sound empirical data.*

Preventing Illness Among People with Coronary Heart Disease, edited by John D. Piette, PhD, Robert M. Kaplan, PhD, and Joseph R. Ferrari, PhD (Vol. 13, No. 1/2, 1996). *"A useful contribution to the interaction of physical health, mental health, and the behavioral interventions for patients with CHD." (Public Health: The Journal of the Society of Public Health)*

Sexual Assault and Abuse: Sociocultural Context of Prevention, edited by Carolyn F. Swift, PhD* (Vol. 12, No. 2, 1995). *"Delivers a cornucopia for all who are concerned with the primary prevention of these damaging and degrading acts." (George J. McCall, PhD, Professor of Sociology and Public Administration, University of Missouri)*

International Approaches to Prevention in Mental Health and Human Services, edited by Robert E. Hess, PhD, and Wolfgang Stark* (Vol. 12, No. 1, 1995). *Increases knowledge of prevention strategies from around the world.*

Self-Help and Mutual Aid Groups: International and Multicultural Perspectives, edited by Francine Lavoie, PhD, Thomasina Borkman, PhD, and Benjamin Gidron* (Vol. 11, No. 1/2, 1995). *"A helpful orientation and overview, as well as useful data and methodological suggestions." (International Journal of Group Psychotherapy)*

Prevention and School Transitions, edited by Leonard A. Jason, PhD, Karen E. Danner, and Karen S. Kurasaki, MA* (Vol. 10, No. 2, 1994). *"A collection of studies by leading ecological and systems-oriented theorists in the area of school transitions, describing the stressors, personal resources available, and coping strategies among different groups of children and adolescents undergoing school transitions." (Reference & Research Book News)*

Religion and Prevention in Mental Health: Research, Vision, and Action, edited by Kenneth I. Pargament, PhD, Kenneth I. Maton, PhD, and Robert E. Hess, PhD* (Vol. 9, No. 2 & Vol. 10, No. 1, 1992). *"The authors provide an admirable framework for considering the important, yet often overlooked, differences in theological perspectives." (Family Relations)*

Families as Nurturing Systems: Support Across the Life Span, edited by Donald G. Unger, PhD, and Douglas R. Powell, PhD* (Vol. 9, No. 1, 1991). *"A useful book for anyone thinking about alternative ways of delivering a mental health service." (British Journal of Psychiatry)*

Ethical Implications of Primary Prevention, edited by Gloria B. Levin, PhD, and Edison J. Trickett, PhD* (Vol. 8, No. 2, 1991). *"A thoughtful and thought-provoking summary of ethical issues related to intervention programs and community research." (Betty Tableman, MPA, Director, Division. of Prevention Services and Demonstration Projects, Michigan Department of Mental Health, Lansing)* Here is the first systematic and focused treatment of the ethical implications of primary prevention practice and research.

Career Stress in Changing Times, edited by James Campbell Quick, PhD, MBA, Robert E. Hess, PhD, Jared Hermalin, PhD, and Jonathan D. Quick, MD* (Vol. 8, No. 1, 1990). *"A well-organized book. . . . It deals with planning a career and career changes and the stresses involved." (American Association of Psychiatric Administrators)*

Prevention in Community Mental Health Centers, edited by Robert E. Hess, PhD, and John Morgan, PhD* (Vol. 7, No. 2, 1990). *"A fascinating bird's-eye view of six significant programs of preventive care which have survived the rise and fall of preventive psychiatry in the U.S." (British Journal of Psychiatry)*

Protecting the Children: Strategies for Optimizing Emotional and Behavioral Development, edited by Raymond P. Lorion, PhD* (Vol. 7, No. 1, 1990). *"This is a masterfully conceptualized and edited volume presenting theory-driven, empirically based, developmentally oriented prevention." (Michael C. Roberts, PhD, Professor of Psychology, The University of Alabama)*

The National Mental Health Association: Eighty Years of Involvement in the Field of Prevention, edited by Robert E. Hess, PhD, and Jean DeLeon, PhD* (Vol. 6, No. 2, 1989). *"As a family life educator interested in both the history of the field, current efforts, and especially the evaluation of programs, I find this book quite interesting. I enjoyed reviewing it and believe that I will return to it many times. It is also a book I will recommend to students." (Family Relations)*

A Guide to Conducting Prevention Research in the Community: First Steps, by James G. Kelly, PhD, Nancy Dassoff, PhD, Ira Levin, PhD, Janice Schreckengost, MA, AB, Stephen P. Stelzner, PhD, and B. Eileen Altman, PhD* (Vol. 6, No. 1, 1989). *"An invaluable compendium for the prevention practitioner, as well as the researcher, laying out the essentials for developing effective prevention programs in the community. . . . This is a book which should be in the prevention practitioner's library, to read, re-read, and ponder." (The Community Psychologist)*

Prevention: Toward a Multidisciplinary Approach, edited by Leonard A. Jason, PhD, Robert D. Felner, PhD, John N. Moritsugu, PhD, and Robert E. Hess, PhD* (Vol. 5, No. 2, 1987). *"Will not only be of intellectual value to the professional but also to students in courses aimed at presenting a refreshingly comprehensive picture of the conceptual and practical relationships between community and prevention." (Seymour B. Sarason, Associate Professor of Psychology, Yale University)*

Prevention and Health: Directions for Policy and Practice, edited by Alfred H. Katz, PhD, Jared A. Hermalin, PhD, and Robert E. Hess, PhD* (Vol. 5, No. 1, 1987). *Read about the most current efforts being undertaken to promote better health.*

The Ecology of Prevention: Illustrating Mental Health Consultation, edited by James G. Kelly, PhD, and Robert E. Hess, PhD* (Vol. 4, No. 3/4, 1987). *"Will provide the consultant with a very useful framework and the student with an appreciation for the time and commitment necessary to bring about lasting changes of a preventive nature." (The Community Psychologist)*

Beyond the Individual: Environmental Approaches and Prevention, edited by Abraham Wandersman, PhD, and Robert E. Hess, PhD* (Vol. 4, No. 1/2, 1985). *"This excellent book has immediate appeal for those involved with environmental psychology . . . likely to be of great interest to those working in the areas of community psychology, planning, and design." (Australian Journal of Psychology)*

Prevention: The Michigan Experience, edited by Betty Tableman, MPA, and Robert E. Hess, PhD* (Vol. 3, No. 4, 1985). *An in-depth look at one state's outstanding prevention programs.*

Studies in Empowerment: Steps Toward Understanding and Action, edited by Julian Rappaport, Carolyn Swift, and Robert E. Hess, PhD* (Vol. 3, No. 2/3, 1984). *"Provides diverse applications of the empowerment model to the promotion of mental health and the prevention of mental illness." (Prevention Forum Newsline)*

Monographs "Separates" list continued at the back

Understanding Ecological Programming: Merging Theory, Research, and Practice

Susan Scherffius Jakes
Craig C. Brookins
Editors

Understanding Ecological Programming: Merging Theory, Research, and Practice has been co-published simultaneously as *Journal of Prevention & Intervention in the Community*, Volume 27, Number 2 2004.

The Haworth Press, Inc.

New York • London • Victoria (AU)
www.HaworthPress.com

Understanding Ecological Programming: Merging Theory, Research, and Practice has been co-published simultaneously as *Journal of Prevention & Intervention in the Community*™, Volume 27, Number 2 2004.

The development, preparation, and publication of this work has been undertaken with great care. However, the publisher, employees, editors, and agents of The Haworth Press and all imprints of The Haworth Press, Inc., including The Haworth Medical Press® and Pharmaceutical Products Press®, are not responsible for any errors contained herein or for consequences that may ensue from use of materials or information contained in this work. Opinions expressed by the author(s) are not necessarily those of The Haworth Press, Inc. With regard to case studies, identities and circumstances of individuals discussed herein have been changed to protect confidentiality. Any resemblance to actual persons, living or dead, is entirely coincidental.

The Haworth Press, Inc., 10 Alice Street, Binghamton, NY 13904-1580 USA

Cover design by Lora Wiggins

Library of Congress Cataloging-in-Publication Data

Understanding ecological programming : merging theory, research and practice / Susan Scherffius Jakes, Craig C. Brookins, editors.
     p. cm.
    "Co-published simultaneously as Journal of Prevention & Intervention in the Community, Volume 27, Number 2 2004."
    Includes bibliographical references and index.
    ISBN 0-7890-2458-6 (hard cover : alk. paper) – ISBN 0-7890-2459-4 (soft cover : alk. paper)
    1. Community health services. 2. Community health services–Citizen participation. 3. Coalition (Social sciences) 4. Health planning–Citizen participation. 5. Substance abuse–Prevention. 6. Health promotion. 7. Regional medical programs. I. Jakes, Susan Scherffius. II. Brookins, Craig C. III. Journal of Prevention & Intervention in the Community.
 RA445.U46 2004
 362.12–dc22

2004001422

Indexing, Abstracting & Website/Internet Coverage

Journal of Prevention & Intervention in the Community

This section provides you with a list of major indexing & abstracting services. That is to say, each service began covering this periodical during the year noted in the right column. Most Websites which are listed below have indicated that they will either post, disseminate, compile, archive, cite or alert their own Website users with research-based content from this work. (This list is as current as the copyright date of this publication.)

Abstracting, Website/Indexing Coverage . . . . Year When Coverage Began

- *Behavioral Medicine Abstracts* **1996**
- *CAB ABSTRACTS c/o CAB International/CAB ACCESS* . . . .
  available in print, diskettes updated weekely, and on
  INTERNET. Providing full bibliographic listings, author
  affiliation, augmented keyword searching.
  <http://www.cabi.org/> **2004**
- *CINAHL (Cumulative Index to Nursing & Allied Health*
  Literature), in print, EBSCO, and Silverplatter, Data-Star,
  and PaperChase. (Support materials include Subject
  Heading List, Database Search Guide, and instructional
  video). <http://www.cinahl.com> **2003**
- *CNPIEC Reference Guide: Chinese National Directory*
  of Foreign Periodicals **1996**
- *Educational Research Abstracts (ERA) (online database)*
  <http://www.tandf.co.uk/era> **2002**
- *EMBASE/Excerpta Medica Secondary Publishing*
  Division <http://www.elsevier.nl> **1996**
- *e-psyche, LLC <http://www.e-psyche.net>* **2001**
- *Family & Society Studies Worldwide <http://www.nisc.com>* **1996**

(continued)

(continued)

Special Bibliographic Notes related to special journal issues (separates) and indexing/abstracting:

- indexing/abstracting services in this list will also cover material in any "separate" that is co-published simultaneously with Haworth's special thematic journal issue or DocuSerial. Indexing/abstracting usually covers material at the article/chapter level.
- monographic co-editions are intended for either non-subscribers or libraries which intend to purchase a second copy for their circulating collections.
- monographic co-editions are reported to all jobbers/wholesalers/approval plans. The source journal is listed as the "series" to assist the prevention of duplicate purchasing in the same manner utilized for books-in-series.
- to facilitate user/access services all indexing/abstracting services are encouraged to utilize the co-indexing entry note indicated at the bottom of the first page of each article/chapter/contribution.
- this is intended to assist a library user of any reference tool (whether print, electronic, online, or CD-ROM) to locate the monographic version if the library has purchased this version but not a subscription to the source journal.
- individual articles/chapters in any Haworth publication are also available through the Haworth Document Delivery Service (HDDS).

ABOUT THE EDITORS

Susan Scherffius Jakes, PhD, is an Extension Specialist serving as the Community Editor of the National Extension website, the Children, Youth, and Families Education and Research Network (http://www. CYFERNet.org) and Director for the North Carolina State University. Her research focuses on program design and evaluation of community-based educational programs and evaluation of community development. The goal of this work is to partner with local level change agents in seeking to infuse their prevention programs with concepts from the ecological model. Dr. Jakes has authored and co-authored articles and web publications on issues related to community development, program design and evaluation, and dissemination. Her formal training is in Community Psychology with an emphasis in evaluation and systems change.

Craig C. Brookins, PhD, is Associate Professor of Psychology and Multidisciplinary Studies and Director of Africana Studies at North Carolina State University. He received his BS from Bradley University in Peoria, Illinois and his MS and PhD from Michigan State University in East Lansing, Michigan where he specialized in Ecological/Community Psychology. Dr. Brookins has conducted research and published in the areas of child abuse and neglect prevention, youth empowerment, African American self-concept and identity development and promotion, community and school-based intervention, independent Black Schools; and adolescent pregnancy prevention and parenting. Over the past 12 years he has also been involved in the development and implementation of a variety of community-based rites-of-passage programs in the Research Triangle Area of North Carolina. Dr. Brookins has provided research consultation and technical assistance to several community-based programs arising out of both public and private not-for-profit agencies. His community work has earned him an *Alumni Distinguished Away for Extension* and he is a charter member of the *Academy of Outstanding Faculty Engaged in Extension* at North Carolina State University.

Understanding Ecological Programming: Merging Theory, Research, and Practice

CONTENTS

Introduction:
Understanding Ecological Programming:
Merging Theory, Research, and Practice

Susan Scherffius Jakes
Craig C. Brookins

North Carolina State University

SUMMARY. In this collection, we examine the application of the ecological model in prevention programs. A review of the literature presents a historical account of the development of the model and provides a basis for the rationale behind its use. Four empirical articles then provide a method for measuring the application of an ecological framework in program design and implementation, and evaluations of programs using components of the model. We conclude that ecological programming, while popular in rhetoric, is complex in both its design and implementation. Ecological approaches show merit over one-sided solutions, but

Susan Scherffius Jakes, PhD, is an Extension Specialist, Department of Family and Consumer Sciences, Box 7605 North Carolina State University, Raleigh, NC 27695 (E-mail: susan_jakes@ncsu.edu).

Craig C. Brookins, PhD, is Associate Professor, in the Department of Psychology, Box 7801, North Carolina State University, Raleigh, NC 27695 (E-mail: craig_brookins@ncsu.edu).

Address correspondence to Susan Jakes.

This manuscript is based on the first author's doctoral dissertation work.

[Haworth co-indexing entry note]: "Introduction: Understanding Ecological Programming: Merging Theory, Research, and Practice." Jakes, Susan Scherffius, and Craig C. Brookins. Co-published simultaneously in *Journal of Prevention & Intervention in the Community* (The Haworth Press, Inc.) Vol. 27, No. 2, 2004, pp. 1-11; and: *Understanding Ecological Programming: Merging Theory, Research, and Practice* (ed: Susan Scherffius Jakes, and Craig C. Brookins) The Haworth Press, Inc., 2004, pp. 1-11. Single or multiple copies of this article are available for a fee from The Haworth Document Delivery Service [1-800-HAWORTH, 9:00 a.m. - 5:00 p.m. (EST). E-mail address: docdelivery@haworthpress.com].

1

need further investigation to show when these are the most efficacious approach and when only limited adherence is optimal. *[Article copies available for a fee from The Haworth Document Delivery Service: 1-800-HAWORTH. E-mail address: <docdelivery@haworthpress.com> Website: <http://www.HaworthPress.com> © 2004 by The Haworth Press, Inc. All rights reserved.]*

KEYWORDS. Ecological programming, evaluation, social problems, prevention

As we work to increase the ability of social programs to effect positive change, we need a better understanding of what the research and theory on ecological programming really say for the program developer. From this we can develop a heuristic for looking into the black box of social programs and understand where a particular program lies on the continuum of adherence to an ecological programming framework. By better defining an ecological model for social programming, we can improve both (a) program developers' awareness of the components necessary to adhere to this model and (b) the ability of future research to measure ecological programming and investigate its predictors and outcomes.

The idea for this special issue on ecological programming came from a recent collaborative experience between university researchers and community-based interventionists. As program developers, we are often charged with designing ecological prevention programs and working with local agents to implement them in communities. Never afraid of asking a dumb question and wanting to be sure all participants were thinking about the same thing, the first author queried during one of these meetings, "What is an ecological program?" The responses from around the room were quite diverse and it was unclear whether there was an essential component that people were thinking of or a collection of many things. This lack of clarity on the part of the researchers and practitioners pointed to a clear gap in our understanding of bringing together theory and practice with regard to ecological programming. For example; What does it mean for a program to be ecological? What do real programs that have implemented these principles look like? How realistic is it to suggest that one should implement an ecological program; is it harder than it seems?

The articles in this publication begin to answer some of these questions by providing a brief historical review of ecological approaches to programming; describing the development of an instrument to operationalize an ecological programming model; and presenting empirical studies of programs using ecological or similar approaches. The goal is to provide the reader with a more in-depth understanding of ecological programming than what currently exists. It is our hope that this understanding can later be used in program development research that tests program models against outcomes to begin answering such questions as: What are the components of ecological programming? What are the outcomes of programming with an ecological model? When is an ecological approach the optimal approach? When is it not? What is the cost/benefit ratio of an ecological approach?

We begin here with a history of the application of ecological concepts to social phenomena and extract what that literature says for social program development.

THE HISTORY AND EVOLUTION OF THE ECOLOGICAL MODEL IN SOCIAL PROGRAMMING LITERATURE

Before an ecological perspective was widely recognized, there were innovative theorists who recognized that people's behavior was not only the product of something that arose from the individual self, but was a product of an interaction of the self with the environment. This early writing was represented by Lewin's (1933) theory that saw behavior as a function of the person and the environment $[B = F(P, E)]$, and Murray's (1938) notion of *environmental press* that saw behavior as created by the interaction of the person's needs and his (or her) environment. These ideas were an eventual springboard for the theory and research in environmental and ecological psychology (Barker, 1968; Mischel, 1968).

Lewin's (1933) *Behavior = F(Person, Environment)* was landmark in its description of the person's behavior as a complex interaction of the person and the environment. But the research that came in the next three decades continued to focus more on the individual than the environment in the shaping of behavior. Throughout the 1960s and 1970s, there was a progression of theory and research showing first how the ecological and psychological were inextricably intertwined to later showing how behavior could even be predicted through an understanding of the behavior setting (Proshansky, Ittelson, & Rivlin, 1976; Schoggen, 1989; Stokols, 1977). This is the fundamental concept of the ecological programming model–that people's behavior cannot be separated from its setting, therefore to change behavior, you must not only change the individual behavior, but change the setting in which it occurs (Vincent & Trickett, 1983).

Barker's 1968 book on Behavior Setting theory and measurement is based on the fundamental assumption that people and their settings are in a continual state of reciprocal influence. He states, "From this viewpoint the environment is seen to consist of highly structured, improbable arrangements of objects and events which coerce behavior in accordance with their own dynamic patterning" (1968, p. 4). This dynamic patterning inspired his life's work of assessing behavior settings. Barker's work takes us a step beyond Lewin's conception of the environment as only the post-perceptual life-space, to encompass the whole of the ecological environment, perceived and unperceived. He does this by defining and measuring behavior settings in an effort to show that behavior can often be better predicted from knowledge of the setting than from knowledge of the individual.

This was the beginning of the revolution in psychology that questioned the foundation of all person-centered psychological theory. It suggested that behavior may be better predicted by the characteristics of the particular environment than by the characteristics of the person in that environment (Barker, 1968; Mischel, 1968). As a result there has been a gradual shift in psychological practice from seeking to change individual behavior through only individual-focused intervention, to behavior change through changing the environment in addition to the characteristics of the person.

The Ecological Analogy

Systems theorists have also applied their knowledge to understanding social issues and social change. The ideas of multiple and possibly interrelated causes of social problems prompted a basic questioning of the purely medical model for understanding individual as well as social problems. The idea of using an "ecological" perspective in program development was expanded through Kelly's 1968 *Ecological Analogy.* The ecological analogy states,

> Interrelationships between the functions of social units and the participation of individual members then become a primary focus for designing programs of interventions where the intervention rearranges the interrelationships or couplings between individual behavior and social functions as much as it alters the behavior of

one social unit or the expressive behavior of any one member of society. (Kelly, 1968, p. 76)

This is to say that the behavior of the individual exists within a context. Much of human behavior is an adaptive or maladaptive response to social events and roles in society. Therefore, to effectively change a behavior, the target of intervention must be the behavior-context interaction. Kelly later explains, "Context is not just something; it is the heart and soul of the matter" (1999). This also was a new direction for psychology as it looked beyond Lewin's perceived environment to describe it in terms of the actual nature of contexts. To understand this nature of contexts, or ecology, he looked to biology for an analogy fitting to human behavior-setting interactions.

Integrating the literature from biology with community psychology, Kelly defined four principles of an ecological analogy that included: (a) Interdependence–in a system, changes in one component of the system dynamically result in changes in another component of the system. (b) Cycling of resources–in a system, the cycling of resources through the processes of production, use, and distribution alerts us to whether resources are present, accessible, integrated, and efficiently used. (c) Adaptation–the process of community change often occurs as a reaction to changes in a community's human, financial, and social resources. Moreover, systems move toward homeostasis, or balance among competing forces within the system, through the behavior of members of a community that will change in order to stabilize the community. Sustainability, the long-term maintenance of healthy community structures, is mediated by the resources of the community and the community's ability to adapt. (d) Succession–the cycling of populations in and out of a community that results in a process of ever-changing resource availability.

Kelly's analogy has several implications for social programs (Mills & Kelly, 1972; Vincent & Trickett, 1983). Applying the Principle of Goodness of Fit, survival and adaptation to a community are often issues of how well the person and environment interact, no matter what the characteristics of the person or the environment. As problems arise within the interaction of the person and environment social programs could then target this interaction to make it a better fit. Another important principle of the ecological analogy is the need for attention to issues at multiple levels of analysis. The direct target of an intervention (e.g., an individual, organization, etc.) may not be the only level at which intervention is necessary to change the system. In addition, unintended effects may arise due to changes in any one part of the system causing reactions in multiple layers of the system (Kelly, 1968). As environments change, so do the roles they create. People may then respond with behavior that is adaptive or maladaptive.

In a further elucidation of the inherent complexity of systems, Bronfenbrenner (1979) in his Model of the Human Ecology, introduced the concepts of embedded systems and dynamic interaction. In an embedded system, the interaction of the individuals, groups, or systems on one level impacts all other levels of the system. In a dynamic system, humans are active and shape the environments in which they live over time. Bronfenbrenner suggested several levels at which the individual and the environmental systems interact. This is often pictured as the systems being depicted as concentric circles representing embedded systems from the individual microsystem to the macrosystemic social order and cultural norms that underlie the consistencies in the inner circles (Bronfenbrenner, 1979). The ecological programming paradigm makes the shift from individual-focused interventions to seeking to help the individual by changing the micro-, exo-, meso- and macrosystem environment; "that is, by placing major attention on the creation and maintenance of challenging, supportive, and responsive environments, both proximate and distal" (Whittaker, Schinke, & Gilchrist, 1986, pp. 491-492).

Bronfenbrenner saw his embedded systems theory as a way to illustrate the ability of interventions to make changes in the whole system. Because systems were embedded in each other, events at one level of the system impacted all other levels. So, to change individual behavior, interventions must have an impact on the systemic behavior setting.

Frequently, this call for broader system changes often led to one-sided solutions, in the direction of the environment, that called for a balance to be rearticulated. In his 1981 article "In Praise of Paradox," Rappaport argues that the most efficacious interventions work with both the target individuals and their environments. Most importantly, Rappaport's dialectic and other work in community psychology and health promotion describe the necessity of multifaceted solutions to social problems (Haignere, 1999; Rappaport, 1981; Rappaport, Davidson, Wilson, & Mitchell, 1975).

The work of these theorists brought to light the reciprocal relationship between behavior and environment. Not only does environment shape behavior, but also behavior shapes the environment. To Lewin's work, Kelly's, Bronfenbrenner's, and later Bandura's theory adds the equation *Environment = F(Person, Behavior)*. This sheds new light on the complexity of the behavior-person-environment interaction.

Applying these ideas to program-development practices, many applicable components of "ecological" programs become apparent. Program developers have long been applying concepts from the ecological model to program practice, but a comprehensive application of the theory requires the components and principles to be clearly defined for program development and operationalized for measurement of adherence to the model.

Health Promotion and Ecological Programming

While psychology continues to advance in its applications of the ecological model to social programming, the field of health promotion has also taken the concept and applied it to a wide range of programs and health interventions (Goodman, Wandersman, Chinman, Imm, & Morrissey, 1996; Moos, 1979). Stokols (1992), in an article on social ecological theory and health promotion, gives four core assumptions of the field. (a) Health is influenced by multiple facets of the physical and social environment. (b) Health promotion and research must address the complex and multidimensional nature of human ecology. (c) Social-ecological health analyses must use multiple methods to address multiple levels of analysis. (d) The social-ecological perspective incorporates the systems theory components of interdependence, homeostasis and negative feedback. From this field we also see continued development of the thinking on what kinds of environments interact with health behavior. Winett (1985) delineates health-determining environments in several of Bronfenbrenner's systems of the human ecology: the interpersonal environment, the informational environment, the ecology of cities, economic influences, regulatory influences, and political-ideological contexts. The literature in health promotion certainly advocates for an ecological approach although the clear delineation of key individual program characteristics remains elusive here as well.

Ecological Applications

Since these many theorists' writings and subsequent research supporting their work, the structure of interventions has not matched the research-based paradigm shift on what works. A study of adolescents that illustrates this perspective states, "It seems likely that the disparity between the problem definition and the focus of intervention is related to the ineffectiveness of traditional services for adolescents; while the intervention focuses mainly on internal dimensions of the adolescent, the relationship between the adolescent and the various social systems–educational, legal, mental health–is virtually ignored" (Kraft & DeMaio, 1982, p. 132). There are, however, program developers focused on understanding the ecological context of social problems and developing interventions from this framework.

This collection seeks to highlight ecological approaches in intervention and evaluation. All four articles are undergirded by the ecological programming model in its application and assessment. Each also highlights different problem areas and a different ecological component or issue.

The Jakes article seeks to expand and clarify the notion of an ecological program. The Ecological Programming Scale (EPS) is presented, analyzed, and compared to other methods for assessing the degree to which a program is ecological. The EPS components load in 3 factors and a single variable: Factor (1) the community control of the program, Factor (2) the systemic change focus of the program, Factor (3) the degree to which the program is multidimensional, and Variable (1) the degree to which the program components are integrated. The single EPS integration variable was the strongest predictor of the experts' ratings of the degree to which a program was ecological. This study provides a compilation of what the literature suggests are the necessary components of an ecological approach and then presents a contrast with what program developers consider to be ecological. This expanded definition is useful for program designers in realizing the possibilities of more fully implementing this approach.

The Smith, Wolf, Cantillon, Thomas, and Davidson article highlights the evaluation of a new iteration of the adolescent diversion project. This project exemplifies an intervention designed with the principles of ecological theory in mind. It uses multiple levels of intervention, working within the youth's natural environment, to reduce the recidivism rates of juveniles that have been caught in delinquent acts. The many iterations of the research study help to tease out the importance of events in the juveniles' environment as a predictor of behavior and how these behaviors can be changed through ecological intervention.

Perkins and Borden evaluate a community youth development program–the Youth Action Program (YAP). The ecological dimensions of YAP were examined through a multidimensional evaluation design and a post-hoc analyses using the components of the EPS. The EPS analysis and evaluation reveal gaps in the program's adherence to the ecological model that provide insight into the program's shortcomings and ultimately the reasons why it no longer exists. The authors assert that if the program had a stronger emphasis on developing community control it might have had more sustainability. This is an interesting look into what happens when some components of ecological programming are emphasized (in this case the multidimensional aspect), but not others.

The Mitchell, Stone-Wiggins, Stevenson and Florin article adds an important dimension to this publication in its introduction to the complexity of creating successful ecological interventions. While it does not look directly at the degree to which programs are ecological it does examine the practical challenges faced by interventionists. An important aspect of ecological interventions is developing collaborative efforts in communities and this is often done through coalition-building. Coalitions have the potential to strengthen all three factors of the EPS by giving the community more control, helping to change systemic forces at the community level and, when coupled with microsystem efforts, creating a multidimensional intervention. This study shows that effecting community change via coalitions is not a turn key process, however. Coalitions are organic and highly impacted by political forces, expertise within the group, and the community's readiness for change, to name just a few issues. The provision of technical assistance is often seen as a solution to navigating these complexities. Using findings from key informants, Mitchell et al.'s work suggests that even this solution is often ineffective at creating improvement in the likelihood of success for coalitions. This is particularly the case when other ecological dimensions are not in place such as the presence of diverse community representation within the coalition, effective organization of these diverse participants, or legitimate opportunities for coalition members to contribute to leadership within the coalition.

The goals of this collection are to provide examples that clarify what ecological applications in intervention look like and create a framework for building our understanding of the strengths and complexities of applying an ecological model in program design, implementation, and evaluation. To this end, Jakes' Ecological Programming Scale is clearly a useful tool that can be used by preventionists to expand the realm of possibilities for program design and assess the "ecologicalness" of their programs. The Smith et al. study, in particular, also confirms the value and role of grounded psychological theory to program development and its importance to the success of an ecological approach.

Finally, the articles overall point to the fact that ecological programming is a process embedded within the dynamics of communities that reflects itself as an art as much as it does a science. This process is not a static one but instead unfolds within constantly changing contexts that are difficult to predict. This, of course, has implications not only for prevention programs but their subsequent replication and dissemination. This latter point is perhaps most sobering for change agents engaged with intractable social problems.

REFERENCES

Bandura, A. (1986). *Social foundations of thought and action: A social cognitive theory.* Englewood Cliffs, NJ: Prentice-Hall.

Barker, R. G. (1968). *Ecological psychology: Concepts and methods for studying the environment of human behavior.* Stanford, CA: Stanford University Press.

Bronfenbrenner, U. (1979). *The ecology of human development.* Cambridge, MA: Harvard University Press.

Bronfenbrenner, U. (1995). Developmental ecology though space and time: A future perspective. In P. Moen, G. Elder, & K. Luescher (Eds.), *Examining lives in context: Perspectives on the ecology of human development* (pp. 619-647). Washington, DC: American Psychological Association.

Goodman, R. M., Wandersman, A., Chinman, M., Imm, P., & Morrissey, E. (1996). An ecological assessment of community-based interventions for prevention and health promotion: Approaches to measuring community coalitions. *American Journal of Community Psychology, 24*(1), 33-61.

Haignere, C. S. (1999). Closing the ecological gap: The private/public dilemma. *Health Education Research, 14*(4), 507-518.

Jakes, S. (2004). Understanding ecological programming: Evaluating program structure through a comprehensive assessment tool. *Journal of Prevention & Intervention in the Community, 27*(2), 13-28.

Kelly, J. G. (1968). Toward an ecological conception of preventative interventions. In J. W. Carter (Ed.), *Research contributions from psychology to community mental health* (pp. 75-99). New York: Behavioral Publications, Inc.

Kelly, J. G. (1999). Contexts and community leadership: Inquiry as an ecological expedition. *American Psychologist, 54*(11), 953-961.

Kraft, S. P., & DeMaio, T. J. (1982). An ecological intervention with adolescents in low-income families. *American Journal of Orthopsychiatry, 52*(1), 131-140.

Lewin, K. (1933). *A Dynamic Theory of Personality: Selected Papers.* New York, NY: McGraw-Hill.

McLeroy, K. R., Bibeau, D., Steckler, A., & Glanz, K. (1988). An ecological perspective on health promotion programs. *Health Education Quarterly, 15*(4), 351-377.

Mills, R. C. & Kelly, J. G. (1972). Cultural adaptation and ecological analogies: Analysis of three Mexican villages. In S. E. Golann & C. Eisdorfer, *Handbook of community mental health* (pp. 157-205). New York: Appleton-Century-Crofts.

Mischel, W. (1968). *Personality and assessment.* New York: Wiley.

Mitchell, R., Stone-Wiggins, B., Stevenson, J. F., & Florin, P. (2004). Cultivating capacity: Outcomes of a statewide support system for prevention coalitions. *Journal of Prevention & Intervention in the Community, 27*(2), 67-87.

Moos, R. H. (1979). Social-ecological perspectives on health. In G. C. Stone, F. Cohen, N. E. Adler & Associates, *Health psychology–a handbook: Theory, applications, and challenges of a psychological approach to the health care system* (pp. 523-547). San Francisco, CA: Jossey-Bass Publishers.

Murray, H. (1938). *Explorations in personality.* New York, NY: Oxford University Press.

Perkins, D. F. & Borden, L. M. (2004). A multidimensional ecological examination of a youth development program for military dependent youth. *Journal of Prevention & Intervention in the Community, 27*(2), 49-65.

Proshansky, H. M., Ittelson, W. H., & Rivlin, L. G. (Eds.). (1976). *Environmental psychology: People and their physical settings* (2nd ed.). New York: Holt, Rinehart, and Winston.

Rappaport, J. (1981). In praise of paradox: A social policy of empowerment over prevention. *American Journal of Community Psychology, 9*, 1-25.

Rappaport, J., Davidson, W. S., Wilson, M. N., & Mitchell, A. (1975). Alternatives to blaming the victim or the environment. *American Psychologist, 30*, 525-528.

Schoggen, P. (1989). *Behavior settings.* Stanford, CA: Stanford University Press.

Smith, E. P., Wolf, A. M., Cantillon, D. M., Thomas, O., & Davidson, W. S. (2004). The Adolescent Diversion Project: 25 years of research on an ecological model of intervention. *Journal of Prevention & Intervention in the Community, 27*(2), 29-47.

Stokols, D. (Ed.). (1977). *Perspectives on environment and behavior: Theory, research, and applications.* New York: Plenum Press.

Stokols, D. (1992). Establishing and maintaining health environments: Toward a psychology of health promotion. *American Psychologist, 47*(1), 6-22.

Vincent, T. A. & Trickett, E. J. (1983). Preventive interventions and the human context: Ecological approaches to environmental assessment and change. In R. D. Felner, L. A. Jason, J. N. Moritsugu, and S. S. Farber (Eds.), *Preventive psychology: Theory, research and practice.* (pp. 67-86). New York: Pergamon Press.

Whittaker, J. K., Schinke, S. P., & Gilchrist, L. D. (1986). The ecological paradigm in child, youth, and family services: Implications for policy and practice. *Social Services Review, 60*, 483-503.

Winett, R. A. (1985). Ecobehavioral assessment in health lifestyles: Concepts and methods. In P. Karoly (Ed.), *Measurement strategies in health psychology.* (pp. 147-182). New York: John Wiley and Sons.

Understanding Ecological Programming: Evaluating Program Structure Through a Comprehensive Assessment Tool

Susan Scherffius Jakes

North Carolina State University

SUMMARY. This study validates the Ecological Programming Scale's (EPS) ability to predict expert ratings of program adherence to an ecological model. Social program descriptions are rated using a previously tested technique and on the EPS. Using principle components factor analysis, the EPS components load on three factors: the community control of the program, the systemic change focus of the program, and the degree to which the program is multidimensional. In comparing the two techniques' prediction of experts' ratings of the degree to which a program is ecological, the single EPS variable measuring program component integration was the strongest predictor. This measurement device may be used to assist researchers in determining the predictors and out-

Susan Scherffius Jakes, PhD, is an Extension Specialist for Child, Family and Community, North Carolina Cooperative Extension, North Carolina State University.

Address correspondence to: Susan Jakes, Department of Family and Consumer Sciences, Box 7605 North Carolina State University, Raleigh, NC 27695 (E-mail: susan_jakes@ncsu.edu).

The author would like to acknowledge the contribution of her mentor and committee chair in this process, Craig C. Brookins. This manuscript is based on the author's doctoral dissertation work.

[Haworth co-indexing entry note]: "Understanding Ecological Programming: Evaluating Program Structure Through a Comprehensive Assessment Tool." Jakes, Susan Scherffius. Co-published simultaneously in *Journal of Prevention & Intervention in the Community* (The Haworth Press, Inc.) Vol. 27, No. 2, 2004, pp. 13-28; and: *Understanding Ecological Programming: Merging Theory, Research, and Practice* (ed: Susan Scherffius Jakes, and Craig C. Brookins) The Haworth Press, Inc., 2004, pp. 13-28. Single or multiple copies of this article are available for a fee from The Haworth Document Delivery Service [1-800-HAWORTH, 9:00 a.m. - 5:00 p.m. (EST). E-mail address: docdelivery@haworthpress.com].

Digital Object Identifier: 10.1300/J005v27n02_02

comes of ecological programming, as well as to provide a more precise framework for developing ecological interventions. *[Article copies available for a fee from The Haworth Document Delivery Service: 1-800-HAWORTH. E-mail address: <docdelivery@haworthpress.com> Website: <http://www.HaworthPress.com> © 2004 by The Haworth Press, Inc. All rights reserved.]*

KEYWORDS. Ecological programming, evaluation, ecological model, community change, social change, program development

Rarely do you hear of a social program these days in mental health, public health, prevention, or intervention, that doesn't make claims of being "ecological." What this means ranges from involvement of parents in a program for children, to the inclusion of a community collaboration component, to efforts designed to change systems at multiple levels. The research-based program development and best practices literature is replete with suggestions on how to make programs ecological, comprehensive, or multi-generational (Dryfoos, 1990; Hobfoll, 1990; Whittaker, Schinke, & Gilchrist, 1986). This research is built on the premise of the efficacy of moving from a problem focus at the individual level of analysis to, more broadly, the greater system of "problems" that need intervention or, even better, prevention. This approach advocates for building on a person's or even a community's strengths or assets. A careful review of this literature, however, leaves one with a nagging confusion given that there remains no clear articulation of what really defines an ecological program. When is a program ecological, and when is it something else? Is there a range of a program's structural adherence to an ecological model?

In community psychology, there have been frequent discussions of ecological methodologies in social programming (Kelly, 1968; Morrissey, Wandersman, Seybolt, Nation, Crusto, & Davino, 1997; Trickett, 1997; Wolff, 1987). Wolff argues that current psychological change efforts are too heavily weighted in the realm of individual direct remedial intervention and that they need to balance the "cube of intervention possibilities" into the areas that: (a) target a higher level of the environment–societal rather than individual; (b) have a higher purpose–empowerment rather than remedial; and (c) use indirect rather than direct methods of intervention (Wolff, 1987).

Within the area of health promotion, there has been an evolution of the operationalization of ecological programming taking place. It began with programs being quantified in terms of the audience. In their 1988 article, McLeroy, Bibeau, Steckler, and Glanz use Bronfenbrenner's Model of Human Ecology to suggest health promotion programs must incorporate programmatic components at multiple levels of the target's ecology. Green and Richard's work built on this framework of audience-intervention diversification and environmental intervention. Their main argument was that McLeroy et al.'s model confused setting and target. An intervention can happen in the organization, but the main targets for change were the individuals, not the organization itself. To remedy this, Richard et al.'s 1996 work differentiated programs by targets, setting, and intervention strategy. This allowed programs to directly or indirectly intervene at multiple levels of the target's ecology in multiple settings. Their work creates a single dimension for comparing programs based on the following criteria. A program is more ecological the more it: (a) integrates environmental and individual targets across a variety of settings; (b) uses at least two different strategies, one with the individual as the direct target and one that targets a component of the environment; and (c) gives more weight to the number of targets of intervention than to the number of settings. They then assigned points to different programs, with more points denoting more adhesion to an ecological framework. This scoring technique allowed programs to be easily compared on their adherence to an ecological model, but the simplicity of the scoring produces severe limitations on the ability to gain an understanding of the degree to which the program is able to affect the ecology of the individual.

The Ecological Programming Scale (EPS) is designed to address these limitations by using the theory on ecological applications to social program development. It sacrifices some simplicity to gain a clearer understanding of the program's ability to impact the ecological context. The EPS assesses the degree to which the structures of social programs adhere to an ecological model. A program's structure is what makes it uniquely able to effect change in its target. Structural components include elements such as program audience, setting, activities, purpose, methods of decision making, underlying theory, etc. Within these structures there exist many options for the types of programs constructed to address a certain problem. The options chosen depend on the underlying theories about the source of the problem and the nature of change. An ecological framework suggests that the source of the problem often occurs outside of the individual and the mechanisms of change not only reduce the source of the problem but also target the dynamic interaction between the individual and the larger system. (See Table 1 for a listing of the components of the EPS.)

Wolff and Richard's models add greatly to the field of ecological program development and evaluation. Although written for different purposes, they share the target and method dimensions of programming. The EPS contains modifications and additions to their components of ecological programming structure that are important in distinguishing programs based on adherence to an ecological framework. These components are (a) an expansion of Richard's intervention setting component to include the degree to which the setting is natural to the target; (b) an adaptation of Wolff's concept of program purpose to include the concept of community empowerment; (c) a measurement of the amount of control the community has over program decisions; (d) a measurement of program adaptiveness, the degree to which a program is responsive to a community's changing needs; and (e) a measurement of the degree to which the program components are integrated in a coherent fashion.

This study, using project descriptions of social programs, explores the underlying factors that contribute to ecological programming as measured by the EPS. It then compares the ability of EPS factors to predict the experts' ratings of the degree to which the programs are ecological over and above that which is predicted using the Richard et al. rating technique. This comparison tests whether the EPS expands the current state of the art of assessing adherence to an ecological model with the following specific research questions: (a) Does the Ecological Programming Scale (EPS) measure ecological programming? and (b) Are the components of the EPS able to predict the experts' ratings of the program as "Ecological" over and above that which can be predicted by Richard's technique alone?

TABLE 1. Components of Ecological Interventions

Factors	Program Components
Multidimensional Programming	Program Audience Program Strategy
Systemic Change	Implementation Setting Target of Change Process of Change Purpose of Program
Community Control	Community Control of Program Funding Community Control of Program Development Community Control of Program Implementation Adaptiveness of Program
Integration Variable	Degree of Component Integration

METHOD

This study, using a program level of analysis, compared 50 social programs developed through the USDA Children, Youth, and Families At-Risk (CYFAR) Initiative. Over the past few years, USDA has put resources into strengthening Cooperative Extension's ability to program effectively with traditionally underserved audiences, including low socioeconomic status groups, ethnic minorities, urban populations and emerging underserved populations. There are currently state-level program year-end reports available for the 300 county programs in 49 states and two territories.

Measures

All programs were rated on the EPS, given a score using Richard's scoring technique, and rated by an expert. Richard, Potvin, Kischuk, Prolic, and Green's (1996) technique assigns program points of 0 to 4 as follows: 0 = single strategy; 1 = two intervention strategies that do not include the direct targeting of the client; 2 = single setting with two strategies in which the individual was the direct target in one; 3 = two settings in which two strategies were implemented with one directly targeting the individual; and 4 = three or more settings in which two strategies were implemented with one directly targeting the individual.

The EPS contains 12 dimensions on which the programs were rated. For each item, a program was scored on the level of the dimension to which it most closely adhered. For example, the degree of program component integration is scored as follows: 0 = there are not multiple components; 1 = the components show little integration, seems like a patchwork of services; 2 = some integration–the program has focus, but not always coherent; and 3 = components are combined in a coherent, integrated fashion. The community control variables, assessed separately for the three specific program periods of obtaining program funding, program design, and program implementation were scored as follows: 0 = Non-participation–Informing; 1 = Consultation; 2 = Community representation; 3 = Partnership-Delegated power; and 4 = Citizen Control (Components developed from the work of Arnstein, 1969). See Table 1 for a listing of the additional components of the EPS.

Once these scores were completed, project descriptions were rated using a 5 question scale assessing the degree to which the program is (a) ecological, (b) likely to produce long term change, (c) comprehensive,

(d) likely to meet community's needs, and (e) exemplary. The questions were anchored on a five-point Likert scale from "not at all" to "very."

Procedure

Sixteen states were randomly selected. From these states, in-depth program descriptions were created based on the 1998 and 1999 project overviews submitted by project directors in their year-end reports. Additional information was obtained from telephone interviews with state-level project directors. If all projects had the same program design within a state, then only one site was assessed in that state; if there was heterogeneity across sites, all sites were used. Completed project descriptions were then rated on the Ecological Programming Scale and scored using the Richard et al. (1996) scoring technique (referred to as the "Richard's score" in the remainder of the study).

At the end of this process, there were 44 programs available for scoring. In order to assure program variability in the level of adherence to an ecological model, Richard's score was used as a simple, previously tested indicator. There were two programs scoring a zero on Richard's score, one scoring one, 13 programs that received a two, 19 scoring three, and nine scoring four. In order to increase the variability in the analysis, six more programs were created that would obtain certain Richard's scores, two each of zeros, ones, and fours. This was done by taking existing programs scoring two or three and removing or adding components as to change the Richard's score. (Only the number of audiences or strategies was changed in order to reduce any bias in the creation of these programs.) These additional programs were also scored on the EPS and given a Richard's score. Twenty-eight percent of the programs were rated twice for interrater reliability testing. Of those programs, with 17 components each, there was an 88% (210/238) agreement in ratings.

Program descriptions were then divided into eight groups, with each group containing programs representing the entire range of Richard's scores. In order to determine interrater convergence on the part of the experts, 14 of the 50 (28%) programs were included in two groups, therefore each group contained 8 programs. These groups of programs were then sent to experts in the field of community-based programming. These experts were university Cooperative Extension faculty from different states working in the area of program development and evaluation and known for being advocates of ecological programming. They were first asked to give brief descriptions or lists of the program components that may increase the following qualities in a program: ecological, comprehensive, likely to produce long-term change, exemplary, and likely to meet community needs. (Due to the space limitations of this article, only the prediction of the variable "ecological" will be discussed.) They were then asked to read all 8 program descriptions and score the programs, using a 5-point scale, on the degree to which they adhered to these characteristics.

The second stage of interrater reliability assessed convergence in the experts' ratings. Instead of being given definitions of the terms on which to rate the programs, the raters were to provide their own definitions of the terms. This highly individualized rating method would therefore produce a lower expected interrater reliability. Because of the nature of the ratings, acceptable convergence was calculated as to be within one unit of agreement. Of the total components, there was perfect agreement on 50% of the ratings, and agreement within one unit on 87% (61/70) of the ratings.

RESULTS

Does the Ecological Programming Scale (EPS) Measure Ecological Programming?

To answer this question, it first had to be assessed whether or not there were underlying dimensions of the EPS. A factor analysis was used to gain a better understanding of the underlying dimensions of ecological programming and to group the intercorrelated components of the EPS into their underlying factors for future analysis. In this analysis, principle components extraction and Promax oblique rotation produced three factors that explained 81.4% of the variance and that all had eigenvalues greater than one (see Table 2).

The three factors are represented in Table 1. There are three intercorrelated components to the EPS. The first factor is the community's influence on programmatic decision making. This will be called *community control* in the remainder of the study. This factor contains the three variables measuring the degree to which community members are given power in the decision making process and one variable measuring the degree to which the program gets feedback and has mechanisms for adapting to the changing needs of the community. The second factor is the degree to which the program works to change the broader system rather than focusing on changing individuals. This will be called *sys-*

TABLE 2. Unrotated Factor Loadings from Principal Components Analysis: Communalities, Eigenvalues, and Percentages of Variance

Item	Component Loading			Communality
	1	2	3	
Program Audience	0.6	0.75	0.05	0.93
Implementation Setting	0.42	−0.56	0.38	0.64
Target of Change	0.76	0.03	0.41	0.74
Process of Change	0.7	−0.13	0.53	0.8
Purpose of Program	0.76	−0.14	0.23	0.65
Program Strategy	0.56	0.79	0.08	0.94
Community Control of Program Funding	0.86	−0.16	−0.41	0.93
Community Control of Program Development	0.84	−0.2	−0.29	0.83
Community Control of Prog. Implementation	0.88	−0.17	−0.38	0.96
ADAPTIVE	0.83	−0.03	−0.2	0.73
Eigenvalues	5.405	1.634	1.103	
% of Variance	54.052	16.343	11.029	

temic change in the remainder of the analysis. It is strongly related to the first factor with a correlation of .589. It contains four variables: (a) the degree to which the program works within existing structures in the community rather than creating new structures, (b) highest system level the program targets in Bronfenbrenner's Model of Human Ecology, (c) the degree to which the program seeks to change the social regularities of the system, and (d) the degree to which the program has a systemic community empowerment focus rather than a remedial focus. The third factor which is moderately related to the first and second (.397 and .327, respectively) is strongly related to the previous methods for measuring ecological programming by Green et al. (1996). It has two primary components, the number of audiences the program targets and the number of strategies the program uses. This will be called *multidimensional programming* in the remainder of the analysis. These three factors all make up important aspects of ecological programming. Also included in the EPS, but not included in the factor analysis due to being a single indicator of a component, is the variable measuring the degree to which the components of the program are integrated in a coherent fashion (called *integration*). This variable was added to the scale as important in getting a more complete picture of the program, but is conceptually distinct from and not significantly correlated with the 3 main factors.

The bivariate regression scores in Table 3 show that there are significant bivariate effects for Factor 3–multidimensional programming ($R^2 = .132$, $p < .01$), and the integration variable ($R^2 = .171$, $p < .01$). These were then entered into the multiple regression analysis to predict the experts' rating of the program as ecological. The overall main effect is significant $F(2, 47) = 6.798$, $p < .05$ but only the partial regression coefficient of integration is significant t $(49) = 2.368$, $p < .05$. This is apparently due to the strong multicollinearity between the independent variables. To test which model is the best fit for predicting the degree to which the experts rated the programs as ecological, an incremental F test was done. This compared the multivariate model of integration and multidimensional programming with the bivariate model of integration as a sole predictor. This nested model compares the multiple correlation squared of the bivariate restricted model with that of the unrestricted model to see if the addition of multiple independent variables improves the prediction ability of the model over and above that of the bivariate model alone. The addition of Factor 3–multidimensional programming–does not add significantly to the ability of the bivariate model of integration to predict the expert's ratings of ecological $F(1, 47) = 3.212$, $p > .05$. It was concluded that the bivariate was the best model.

Does the Variable Measuring the Degree to Which the Components Are Integrated Add to Previous Measures (Richard's et al., 1996) of Ecological Programming?

To address the second research question, the analysis tests the degree to which the components of the EPS add to the prediction of the experts' rating of programs as ecological over and above that which is predicted by Richard's scoring technique alone. Table 3 shows the bivariate regression analysis of Richard's score predicting the degree to which the programs were rated as ecological. As seen in the previous analysis, the best model for the prediction of the experts' rating of ecological from the EPS is the single variable integration. An incremental F test was used to compare the nested model of inclusion of the independent variables integration and Richard's score versus using Richard's score as a sole predictor of the experts' rating of the program as ecological. The

TABLE 3. Summary of Bivariate Regressions for EPS Variables and Richard's Score Predicting the Experts' Rating of the Programs as Ecological

Variables	B	SEB	Beta	R Square
F1 - Community Control	.006	.037	0.232	.054
F2 - Systemic Change	.040	0.043	.133	.054
F3 - Multidimensional Programming	.101	.037	.363**	.132
Integration	0.112	0.036	0.413**	.171
Richard's Score	.294	.121	.330*	.109

* $p < .05$ ** $p < .01$

addition of the variable integration does add significantly to the ability of Richard's scoring technique to predict the experts' rating the program as ecological $F(1, 47) = 5.62, p < .05$. In fact, when the incremental F test is run comparing the multivariate model with the bivariate model with integration as the sole predictor, it is the stronger model $F(1, 47) = 7.67, p < .05$. Integration is such a strong predictor of the experts' rating of ecological, that the bivariate model with integration as the sole predictor is still the best model.

DISCUSSION

This study was designed to validate the Ecological Programming Scale (EPS) by testing whether it is able to better predict experts' rating of a program as ecological over and above previous measures of ecological programming.

What Are the Underlying Dimensions of the EPS?

From the factor analysis, there are three clear dimensions of the EPS. These three factors give a robust picture of the dimensions of ecological programming. The variation in the strength of the intercorrelations between the three factors suggests that community control and systemic change are more strongly associated with each other than with multidimensional programming (the number of strategies and audiences). Community control addresses a philosophy in methodology, the community-driven rather than the expert-driven approach to programming. Systemic change addresses a philosophy in purpose, the program must seek to change the larger system or ecology rather than individuals.

Multidimensional programming is different; while community control advocates for community capacity-building and systemic change for real systems change, multidimensional programming assesses only the quantity of the components of the intervention rather than quality or focus. While this may suggest a philosophy of multimethod and comprehensive programming, a closer examination of the multidimensional programming pattern matrix shows that the multidimensional programming factor has weights that only strongly positively affect three of the variables. The positive effects are on the number of audiences and strategies, but it has a negative influence on the Implementation variable, which measures the degree to which the program is implemented into a newly created or unfamiliar environment rather than into familiar or existing environments. The multidimensional programming factor therefore suggests multiple methods, but gives no indication of the focus of those methods on effecting ecological change.

Do the Components of the EPS Predict the Experts' Rating of the Degree to Which the Programs Are Ecological?

The best fitting model for the prediction of the experts' rating of programs as ecological was in the single variable measuring the degree to which the program components were integrated in a coherent fashion. The logic for this is in the nature of the measure. For the program components to be integrated, there have to be multiple components. A score of zero on this dimension is given to single component programs. The subsequent levels measure the degree of component integration, but also suggest the presence of an underlying theory or framework to the program. To be integrated in a coherent fashion, there has to be some apparent underlying program focus that provides this coherency. Thus integration becomes increasingly important as programs become more multifaceted. This suggests that a program is more ecological–or able to affect the ecology of the individual–the more its multiple components work together for a cumulative program impact on more than the single individual target of intervention.

Does the Variable Measuring the Degree to Which the Components Are Integrated Add to Richard's Technique for Measuring the Degree to Which Programs Are Ecological?

The best model for predicting the experts' rating of the degree to which a program is ecological includes only the program component Integration as bivariate predictor variable. The prediction ability of Richard's score does not add significantly to the prediction ability of the measure of program integration. The scoring technique developed by Richard et al. (1996) is purely quantitative. It increases with the number of audiences and strategies, while valuable, it does not evaluate the quality or coherency of the program's growing complexity. As discussed in the previous section, the integration variable begins to capture the concept of program quality, such that a program that scores high using Richard's technique and on the integration variable is not only multifaceted, but also has a coherent focus tying the program components together. It is logical that a program that uses integrated strategies in multiple levels of the target's ecology is more likely to positively impact the ecology and person in a meaningful way. Because the nature of the integration variable also presumes multiple methods, Richard's score is not able to add to the prediction ability of simply assessing the degree to which the program components are integrated. This parsimonious solution greatly strengthens our ability to measure the degree to which a program is rated as ecological by solely assessing the degree to which the program components are integrated.

So What Happened to the Systemic Change Factor?

In the analysis we see very little association of the systemic change factor with the experts' rating of any of the indicators of ecological or exemplary programming. In the literature review and in other discussions by the author it is argued that systemic change is not only an essential aspect of ecological programming, but may be the defining dimension. In the sample of programs used in this study, there was the full range of variability on all of the questions that made up this dimension, so it was not a problem with the program sample. The issue then must lie in the measurement of the dependent variable. The expert raters in this study have many years of experience in program development and evaluation, and were chosen specifically for their combination of focus on a specific target audience and community capacity-building. The experts were given no information about the study and no definitions of the terms on which they rated the programs. They were asked to give definitions of the terms before scoring the programs. The interrater reliabilities showed consistency across experts in their scoring. There is also consistency in the expert definitions given of the dependent variables. These definitions show an emphasis on multidimensional programming. While some of the experts mentioned the need to create environmental or contextual change and be responsive to community needs, interventions targeted at multiple levels of the system seemed to suffice. A distinction is not made in the experts' definitions of the nature of these changes at these higher levels of the system that would be able to differentiate programs that only seek incremental or rebalancing types of change as compared with those that seek to change the social regularities within the system (Seidman, 1988). These intervention types may maintain the status quo rather than fundamentally change the way that individuals or groups interact with their environment.

How Does This Add to the Literature?

This study gives solid evidence of the need to include an assessment of program component integration when measuring ecological programming. Knowing the number of audiences and strategies and ensuring an individual as well as non-individual intervention (the basis for Richard's score) is heightened in its ability to predict the experts' ratings of programs as ecological with the inclusion of this dimension. Programs can have many components and strategies, but seem disjointed if there is not an underlying coherency to all of the dimensions.

In addition, because this study looked at programs that were encouraged by their funders to be multidimensional, they had a much more "ecological" mean Richard's score than previous studies. This allowed investigation of more within-group differences in the higher ranges of ecological programming. For example, in rating programs on their integration of components, there was a subset of programs that showed poor integration but had many components and were strongly driven by the community. These programs developed from the interest of mobilized community members. Often they had many audiences and strategies, but were seemingly disjointed, completely driven by apparent needs in the community. There was no apparent underlying theory of change between the components, but these programs were effective in that the community members were empowered to identify issues and make changes happen in the community as a whole. Although these programs had many strategies, settings, and audiences (thus high Richard's scores), they were rated lower on the degree to which they were ecological by the experts because of this lack of coherent integration.

This study is also a challenge to push the vernacular to expand what is meant by an ecological program. There are many dictates within the literature on programs that work to be "ecological." This research suggests that this may mean more than having program strategies with multiple targets in multiple settings, but that those strategies must be driven by a coherent theory of change. In addition, as one could argue that to be ecological, a program must also be able to meet the community's needs, this research also points to the need to give community members power in driving program decisions and adaptations. This study was not able to show evidence of the need for systemic change focus to predict the experts' current ratings of a program as ecological, but that may be more an indication of the current state of the rhetoric rather than the lack of a need for these types of approaches in social programs.

Future Research

While this study was able to look at programs across a wide variety of content areas for common structural dimensions, it could not allow for the study of programs based on actual program results. Ideally one would test how "ecological" a program was based on tangible evidence of change in the target's ecology. Across a variety of program content areas, these outcomes become incomparable. In addition, the use of experts' ratings has limitations. Experts' perceptions are subjective and again only able to assess programs based on what has been within their experience to work. This causes problems when the reality is that programming practice does not yet contain consistent strength in all of the areas advocated for by the literature on the ecological model as assessed by the Ecological Programming Scale. When the experts haven't had the opportunity to see strongly ecological programs work they are hard to distinguish and define as such.

Lessons for Programming Practice

Arguably equally important in this work is increasing the knowledge of program developers about the implications of the ecological theory for social change efforts. How is ecological programming related to what we know works in prevention programming? Although it has been outside the scope of this study, it is interesting to note the relationship of ecological programming models to what we know works in prevention programming. Morrisey, Wandersman, Seybolt, Nation, Crusto, and Davino (1997) give an excellent summary of what is known to be important in effective prevention programming and how it relates to practice. Their suggested program characteristics sound very familiar: comprehensive, theoretically-based, intensive, tailored to the needs of the participants, culturally-relevant, appropriately-timed, focused on skill development, and containing sufficient follow-up.

It is also important to note that there are different issues that require different targets of intervention. Even though all social problems have causes that are multiply determined, Haignere points out that some health issues may best be prevented by individually targeted programs, whereas others are more publicly weighted in their causation and may require more systemically focused interventions (1999). Program developers must have a strong research-based understanding of when an ecological approach is appropriate or when an individually focused intervention may be the best solution.

It is important to note that there are very real barriers to implementing effective social programs (Jakes, 1997; Morrisey et al., 1997). Saranson (1972) warns against an idealized notion of the creation of settings. New settings come replete with problems of leadership, structure, expectations and external influences. There are real difficulties that occur in any systems change effort. From funding barriers to personality conflicts, programming in the real world is seldom close to the research based exemplars often implemented with idealized funding in cream-of-the-crop sites (see Morrisey et al., 1997, for an in-depth review). In spite of these challenges, innovative communities can design new ways to create real lasting change in the ways things operate.

Lastly, this study offers a challenge to program developers to seek to cause fundamental change in the social regularities of the setting–not just perpetuate the status quo. If a program does not seek to actually change the setting-person interaction, it is not ecological (Seidman, 1988). As more social innovations are developed that are able to create change on this level, the social regularity of the rhetoric of ecological programming may change as well.

REFERENCES

Arnstein, S. R. (1969). A ladder of citizen participation. *American Institute of Planners Journal, 35*, 216-224.

Bronfenbrenner, U. (1979). *The ecology of human development.* Harvard University Press: Cambridge, MA.

Bronfenbrenner, U. (1995). Developmental ecology though space and time: A future perspective. In P. Moen, G. Elder, and K. Luescher (Eds.), *Examining lives in context: Perspectives on the ecology of human development* (pp. 619-647). Washington, DC: American Psychological Association.

Dryfoos, J. G. (1990). *Adolescents at risk: Prevalence and prevention.* Oxford University Press: New York.

Green, L.W., Richard, L., & Potvin, L. (1996). Ecological foundations of health promotion. *American Journal of Health Promotion, 10,* 270-281.

Haignere, C. S. (1999). Closing the ecological gap: The private/public dilemma. *Health Education Research, 14*(4), 507-518.

Hobfoll, S. E. (1990). Person-environment interaction: The question of conceptual validity. In P. Tolan, C. Keys, F. Chertok, and L. Jason (Eds.), *Researching community psychology* (pp. 164-167). Washington, DC: American Psychological Association.

Jakes, S. S. (1997). *Assessing an ecological prevention programming model in Cooperative Extension county programming.* Unpublished master's thesis, North Carolina State University, Raleigh, NC.

Jakes, S. S. (2001). *Understanding ecological programming: Evaluating program structure through a comprehensive assessment tool.* Unpublished doctoral dissertation, North Carolina State University, Raleigh, NC.

Kelly, J. G. (1968). Toward an ecological conception of preventative interventions. In J. W. Carter (Ed.), *Research contributions from psychology to community mental health* (pp. 75-99). New York: Behavioral Publications, Inc.

McLeroy, K. R., Bibeau, D., Steckler, A., & Glanz, K. (1988). An ecological perspective on health promotion programs. *Health Education Quarterly, 15*(4), 351-377.

Morrissey, E., Wandersman, A., Seybolt, D., Nation, M., Crusto, C., & Davino, K. (1997). Toward a framework for bridging the gap between science and prevention: A focus on evaluator and practitioner perspectives. *Evaluation and Program Planning, 20*(3), 367-377.

Richard, L., Potvin, L., Kischuk, N., Prolic, H., & Green, L. W. (1996). Assessment of the integration of the ecological approach in health promotion programs. *American Journal of Health Promotion, 10,* 318-328.

Saranson, S. B. (1972). *The creation of settings and the future societies.* San Francisco, CA: Jossey-Bass, Inc.

Seidman, E. (1988). Back to the future, community psychology: Unfolding a theory of social intervention. *American Journal of Community Psychology, 16*(1), 3-24.

Trickett, E. J. (1997). Ecology and primary prevention: Reflections on a meta-analysis. *American Journal of Community Psychology, 25,* 197-206.

Whittaker, J. K., Schinke, S. P., & Gilchrist, L. D. (1986). The ecological paradigm in child, youth, and family services: Implications for policy and practice. *Social Services Review, 60,* 483-503.

Wolff, T. (1987). Community psychology and empowerment: An activist's insights. *American Journal of Community Psychology, 15,* 151-166.

The Adolescent Diversion Project:
25 Years of Research
on an Ecological Model of Intervention

Emilie Phillips Smith

University of South Carolina

Angela M. Wolf

National Council on Crime and Delinquency

Dan M. Cantillon
Oseela Thomas
William S. Davidson

Michigan State University

Emilie P. Smith, PhD, is Associate Professor, The Pennsylvania State University, Department of Human Development and Family Studies, University Park, PA 16802.

Angela M. Wolf, PhD, is Senior Researcher, National Council on Crime and Delinquency, 1970 Broadway, Suite 500, Oakland, CA 94612 (E-mail: awolf@sf.nccd-crc.org).

Dan M. Cantillon, PhD, is now a Post-Doctoral Research Fellow, at the University of Illinois at Chicago Health Research and Policy Center (MC 275), 850 West Jackson Blvd., Suite 400, Chicago, IL 60607 (E-mail: cantillo@uic.edu).

Oseela Thomas, PhD, and William S. Davidson, PhD, Professor, are affiliated with the Department of Psychology, 135 Snyder Hall, Michigan State University, East Lansing, MI 48824-1117 (E-mail: thomasos@hotmail.com) or (E-mail: davidso7@pilot.msu.edu).

Address correspondence to: Emilie Smith, PhD, Pennsylvania State University, Department of Human Development and Family Studies, University Park, PA 16802.

[Haworth co-indexing entry note]: "The Adolescent Diversion Project: 25 Years of Research on an Ecological Model of Intervention." Smith et al. Co-published simultaneously in *Journal of Prevention & Intervention in the Community* (The Haworth Press, Inc.) Vol. 27, No. 2, 2004 pp. 29-47; and: *Understanding Ecological Programming: Merging Theory, Research, and Practice* (ed: Susan Scherffius Jakes, and Craig C. Brookins) The Haworth Press, Inc., 2004, pp. 29-47. Single or multiple copies of this article are available for a fee from The Haworth Document Delivery Service [1-800-HAWORTH, 9:00 a.m. - 5:00 p.m. (EST). E-mail address: docdelivery@haworthpress.com].

29

SUMMARY. The Adolescent Diversion Project (ADP) is an ecological program that seeks to promote family and community support and divert youth from further potentially labeling contact with the juvenile justice system. This manuscript reviews results from three phases of research that test program efficacy, compare intervention components, and examine staffing models. The model involves trained paraprofessionals who utilize behavioral contracting and community advocacy to help families of delinquent youth. Results are then presented from the fourth phase, the current study that replicates the model and presents an empirical test of the underlying conceptual framework derived from social-interactionist/labeling theory. Again, ADP is found to result in less official delinquency than the "warn and release" or juvenile justice processing conditions. Perceived negative labeling is related to increased delinquency while perceived awareness of youth activities without labeling is related to reduced delinquency. The results highlight the importance of the family, community, and juvenile justice contexts and their reactions to juvenile delinquency. *[Article copies available for a fee from The Haworth Document Delivery Service: 1-800-HAWORTH. E-mail address: <docdelivery@ haworthpress.com> Website: <http://www.HaworthPress.com> © 2004 by The Haworth Press, Inc. All rights reserved.]*

KEYWORDS. Juvenile delinquency, diversion, ecological program, labeling theory, juvenile justice

Diversion is the process of removing youthful offenders from formal contact with the court system by referring them to community-based service provision programs (Eddy & Gribskov, 1998; Gensheimer, Mayer, Gottschalk, & Davidson, 1987; and President's Commission, 1967). The primary goal of the Adolescent Diversion Project (ADP) was to create an alternative approach to deal with delinquency using rigorous scientific evaluation and juvenile justice policy. The purpose of the present article is to: summarize the theoretical background of the model, to provide a description of over 25 years of prior research on ADP, and to describe results from a replication that also tests the underlying theoretical model.

THEORETICAL BACKGROUND

The conceptual premises of the ADP model involves three major theoretical perspectives: social control and bonding, social learning, and social-interactionist theories.

Social control theory originally outlined by Hirschi (1969) highlights the importance of social bonds in preventing delinquent behavior. Attachment to prosocial family members, peers, and adults in the community is thought to lead to social control that reduces adolescent deviant peer associations likely to result in delinquent behavior (Agnew, 1985; Agnew, 1993; Greenberg, 1999).

The social learning theory perspective views delinquency as learned through interactions with family, peers, and others (Akers, 1990; Winfree, Backstrom, & Mays, 1994). Behaviors are reinforced through a variety of mechanisms including the reinforcers and punishers received in the past, present, and those that are anticipated to occur through both informal and formal sanctions (Akers, 1990).

Social interactionist theory suggests that it is the observation and labeling of behavior as delinquent that results in further social interactions that intentionally or unintentionally label the youth as delinquent (Schur, 1973). Based upon the ascribed label, families, juvenile justice staff, and police either fail to interact positively or sanction the youth in such a way that forecloses legitimate avenues of behavior. Court contact, incarcerating youth, or prescribing participation in programming exclusively with delinquent youth are among the well-intentioned ways of foreclosing access to positive non-delinquent peers. This process negatively labels and encapsulates the youth within a delinquent identity. Research based upon labeling theory has found the stigma and social interactions resulting from court involvement are more harmful than beneficial because it tends to perpetuate delinquent behavior (Ageton & Elliott, 1974; Kaplan & Johnson, 1991; Klein, 1986). Effects of court contact have been found not only upon delinquent behavior, but also upon the delinquent values, orientation, and self-perceptions of youth (Ageton & Elliott, 1974; Downs & Rose, 1991; Gray-Ray & Ray, 1990; Matsueda, 1992).

A combination of social-control theory, social-learning theory, and social-interactionist theory form the conceptual framework of ADP. Behaviorally-based family intervention incorporates the premises of both social control and social learning theory. Behaviorally based family intervention has been found to be one of the more successful approaches in enhancing family relationships and in preventing and reducing delinquency (Alexander, Robbins, & Sexton, 2000; Davidson, Redner, Amdur, & Mitchell, 1988; Henggeler, Schoenwald, Borduin, Rowland, & Cunningham, 1998; Lipsey & Wilson, 1998). The ADP framework attends to family and the multiple ecological contexts in young people's lives. ADP seeks to strengthen bonds and attachment to family and prosocial others (social control theory); to help families establish clear behavioral standards, monitoring, and contingencies (social learning theory); and to divert youth from potentially stigmatizing social contexts, such as the juvenile justice system, and build support within their natural communities (Mendel, 2000) to prevent further delinquency.

THE ADOLESCENT DIVERSION MODEL: BACKGROUND AND EARLIER RESEARCH

ADP was originally developed in Champaign-Urbana in the 1970s, a time in which researchers and policy-makers began searching for "what works" (Davidson, Seidman, Rappaport, Berck, Rapp, Rhodes, & Herring, 1977). ADP was replicated in the state of Michigan extensively (Davidson & Johnson, 1987; Davidson et al., 1987; Davidson et al., 1988) and elsewhere (Smith & Meyers, 1994). The various tests of diversion have been supported by a diversity of funding mechanisms including national and state grants and contracts (i.e., NIMH, OJJDP, State of Michigan). Over the past nearly three decades, ADP has undergone several research phases. This section reviews background research on: (1) program efficacy; (2) examination of intervention components; (3) staffing models and lastly (4) presents a detailed description of an ADP replication and an empirical test of the underlying conceptual model. Table 1 summarizes the phases and results to be described in the following sections.

Phase 1: Program Efficacy

In the initial tests of ADP, conducted in a small mid-western town, youth were referred from the juvenile divisions of the two local police departments. Eighty-four percent were male, 67% White, 30% African American, the average age was 14.3 years and generally, the youth had completed 8 years of school. They had been previously arrested an average of 2.16 times for a variety of offenses with the most common offenses being larceny and breaking and entering. Undergraduate volunteers enrolled for two semesters of college credit received six weeks of training focused on behavioral contracting with families and community advocacy efforts to help youth attain educational, recreational, and occupational goals (Davidson & Rapp, 1976; Davidson et al., 1977). Volunteers delivered 14-16 weeks of behavioral contracting and com-

TABLE 1. Summary of Results Across ADP Research Phases

Phase	Purpose	Design	Results
Phase 1: Efficacy	Examine program efficacy	Random assignment of youth to: 1) ADP 2) Treatment-as-usual (juvenile justice processing)	ADP youth had fewer police and court contacts at 1 and 2-year followup.
Phase 2: Staffing	Examine use of various types of volunteers upon program effects	Random assignment to types of volunteers: 1) University 2) Community college 3) Community (non-college) volunteers 4) Treatment-as-usual	All 3 types of volunteers had fewer court petitions than control group. Community volunteers more difficult to recruit.
Phase 3: Program Components	Test the effects of various aspects of programming	Random assignment to condition: 1) Action (A): contracting and advocacy 2) Action Family focus (AF): contracting and family self-advocates from outset 3) Relationship (R): develops relationship with volunteer, no contracting nor advocacy 4) Action Court setting (C): contracting and advocacy from court 5) Attention Placebo (AP): relationship with volunteer with little training/supervision	Youth in A, AF, and R conditions with all had less official delinquency than AC (Court). Court setting and Placebo had more court petitions than other conditions.
Phase 4: Test of Conceptual Processes	Replicate the model, and test the processes of effects suggested by labeling theory.	Youth randomly assigned to: 1) Diversion with services - ADP 2) Diversion with no services - warn and release 3) Treatment-as-usual - juvenile justice processing	ADP youth had significantly fewer court petitions than warn and release or juvenile justice processing. Negative labeling related to increased delinquency. Awareness of youth activities without labeling related to reduced delinquency.

munity advocacy with youth and families for 6-8 hours per week. The last quarter of the intervention was spent helping the families to implement their own behavioral contracts and advocate for themselves, often in terms of school and pre-employment opportunities. At one- and two-year follow-up, youth randomly assigned to the intervention group had significantly less official delinquency, i.e., fewer police contacts, fewer court petitions filed against them and less severe offenses than youth in the treatment-as-usual control group who were handled by the normal juvenile justice processes (Davidson et al., 1977).

Phase 2: Examining Staffing Models

The major goal of Phase 2 was to examine distinctions between the type of volunteer staff and the program impact upon delinquency (Berman & Norton, 1985; Durlak, 1979). The study examined a total of 129 youth who were randomly assigned to volunteers from a large Midwestern university (47 volunteers), community college volunteers (36), and community volunteers (19) who were all similar demographically in terms of gender and race (except the university students were younger and less likely to have families). The students all received college credit while community volunteers did not. There was less success in involving community volunteers than volunteers from the university or community college. They all received training and supervision in the same intervention model, the ADP behavioral contracting and advocacy model. All of the volunteers exhibited efficacy in delivering the model. The university volunteers at $F(1, 120) = 15.21$, $p < .05$, the community college volunteers at $F(1, 120) = 23.24$, $p < .05$, and the community volunteers at $F(1, 120) = 17.78$, $p < .05$ all had significantly fewer court petitions than the treatment-as-usual juvenile justice processing group (Kantrowitz, 1979; Davidson et al., 1988).

Phase 3: Identification of Successful Model Components

The third phase of ADP began to test variations of the original model with 228 youth who were 83% male, 74% White, 26% African American with a mean age of 14.2 years. Most often they were petitioned to court once for larceny (34%) or breaking and entering (24%). Youth were randomly assigned to one of the alternative ADP models described stratified for gender, ethnicity, and order of referral (Emshoff & Blakely, 1983).

The *action condition* was a replication of the original intervention model that used the techniques of behavioral contracting and child advocacy. The *action condition–family focus* was similar to the Action Condition except advocacy activities were conducted only within the family and the family was encouraged to self-advocate from the outset of the program. The *relationship condition* focused upon developing the relationship between the volunteer and youth but did not utilize the behavioral or advocacy models. Training with the volunteers focused upon unconditional regard, communication, and problem-solving with the individual. The *action condition–court setting* differed from the original model in that supervision of the volunteers was conducted at the local juvenile court solely by the juvenile court caseworker. The *attention placebo condition* was constructed to assess the effects of nonspecific attention and therefore provided minimal training and supervision. The intervention involved mainly recreational activities. Youth in the *control condition* were returned to the local juvenile court for processing as usual.

Multivariate analyses of variance exhibited significant differences between some of the conditions and planned comparisons clarified the direction and impact. The significant difference identified was that the action condition, the action condition–family focus, and the relationship condition were statistically superior to the court-processing control condition with less official delinquency at the 30-month follow-up period. The action condition–court setting and the attention placebo condition were similar and did not perform significantly better than the control group on official delinquency, i.e., the number of court petitions (Davidson et al., 1987; Emshoff & Blakely, 1983).

METHOD

Phase 4: The Current Replication Study

In Phase 4, the primary aim was to evaluate the program's effectiveness when replicated in an urban environment. Additionally, the utilization of paid staff as change agents was evaluated to increase potential dissemination of the ADP model to human service agencies interested in replication. Importantly, Phase 4 provided an opportunity to more thoroughly examine the processes underlying the program, particularly those described earlier in this paper and suggested by social control, social-learning, and social interactionist/labeling theory.

Participating Youth

Fifteen youth officers from four different city precincts referred a total of 395 youth. The same eligibility criteria and referral procedures from prior phases were followed. The average age was 14 years; 84% were male; and the majority (65%) was referred for property-related offenses such as breaking and entering, larceny, and auto theft. Characteristics of referred youths were therefore comparable to those from previous ADP studies with the exception of race. Ninety-one percent of the youths were African American compared to 30%, 26%, and 33% participation in previous research on the model.

Paid family workers from a local service agency received two weeks, approximately 80 hours, of intensive training that included an overview of national and local juvenile justice issues as well as the original eight-unit training manual, homework exercises, and role-plays.

Intervention Models

At the time of the referral, project research staff blind to condition conducted an interview with the youth and at least one parent. The participant's rights, as both a juvenile offender and a voluntary participant in an intervention project were reviewed during this interview. It was explained to the youth and their parents that if they agreed to participate they would be randomly assigned, "by chance" to a program condition. After intake and pre-assessment, participants were randomly assigned to one of the following conditions: diversion with services (ADP), diversion without services (warn and release), or treatment-as-usual control (juvenile justice processing).

Diversion with Services Condition. A total of 137 youth were assigned to this condition. The family worker combined the techniques of child advocacy and behavioral contracting in the youth's natural environment. Youth and families were guided in developing behavioral goals, meaningful rewards and sanctions for the youth, and were assisted in assessing important community resources principally to help the youth advance educationally (many of whom were facing/had faced suspension/expulsion). They also helped to identify options for youth development through hobbies, recreation, volunteerism, work, and pre-employment opportunities. The first 12 weeks were termed the active phase and family workers spent 3 hours weekly with the youth and family while providing direct assistance in behavioral contracting and advocacy efforts. During the last four weeks, or the follow-up phase, family workers spent 1.5 hours of assistance in the same areas, but their role was that of a consultant.

Diversion Without Services Condition. Youth (n = 134) in this condition were returned to their parents with no further program or court contact. All referring charges were dismissed. The only additional requirement for the youth and parent(s) were two additional interviews 4 months from the initial intake meeting, and the second follow-up was 12 months from initial referral.

Treatment-as-Usual Control Condition. A total of 124 youths participated in this condition. Individuals in this condition were returned to the court's jurisdiction for traditional processing which resulted in a petition to the juvenile court. Again, parents and youths in this condition participated in the two follow-up interviews.

Measures

Pre-assessment measures included demographics, labeling, self-reported delinquency, and official delinquency. The self-reported delinquency scale, based upon the work of Elliott, Huizinga, and Ageton (1985), was a 29-item measure assessing illegal and delinquent behavior with an internal consistency (alpha) of .81. Information on Official Delinquency was obtained by coders blind to condition who searched the records of 44 law enforcement jurisdictions, the juvenile court and the Law Enforcement Information Network (LEIN) on the number of police contacts, number of court petitions, seriousness of offense, and disposition.

The labeling measures, developed explicitly for this study, examined both perceived delinquent labeling from others and self-labeling. Measures examining others awareness of delinquency and other's expectation of future delinquency examined the number of people knowledgeable of the youth's act(s) as well as the number of people expecting further criminal behavior (alphas of .41 and .45). Because the youth might report that family, teachers, and neighbors differed in their knowledge and expectancies, these scales did not exhibit considerable internal consistency. The youth also provided information on their perceived reputation with others and their self-perceptions (alphas of .87 and .81, respectively).

The Family Relationships measure developed for this study, consisted of eight items, reported by the youth, measuring family communication, time spent with parents, and subjective ratings of family relationships. This scale had an alpha of .72.

RESULTS

The results section examines three important issues: (1) the integrity of random assignment in this naturalistic experiment supporting the internal validity of the study; (2) an empirical examination of program effects upon crime and delinquency; and (3) a test of the ecological theorized processes suggested by social-interactionist and social control/bonding theory. The following sections address these issues.

INTEGRITY OF RANDOM ASSIGNMENT

Participants in the 3 experimental conditions: traditional court processing, diversion with services, and release to parents were compared to ascertain that they were similar demographically prior to the beginning of the program. This is one method for providing a manipulation check of random assignment. The variables examined included gender, race, school enrollment, family structure, youth's employment status, type of referral offense, socio- economic status, age, grade completed, and self-report delinquency. Categorical variables were analyzed using Chi-Square and ANOVA was used to examine continuous variables. No substantial pre-program differences were detected except for a slight difference in socioeconomic status (Table 2). The difference between the groups on school enrollment approached statistical significance at the .05 level. This is potentially due to court-involved youth penetrating the juvenile justice system more deeply and receiving alternative educational arrangements.

PROGRAM EFFECTS

Repeated Measures Analysis of Variance was used to examine effects by time, (pre, post, and follow-up) and by condition (diversion with services, diversion without services, and court processing as usual). As in previous ADP research, there were no significant differences between the three dispositional groups in self-reported delinquency. However, there was a statistically significant time × condition effect upon official delinquency. Diversion with services demonstrated efficacy in decreasing officially recorded recidivism rates when compared to both the diversion without services and court-processed group

TABLE 2. Comparison of Experimental Groups

Variable	χ^2	DF	PROB
Gender	1.95	2	.38
Race	.95	2	.78
Enrolled in School	5.33	2	.07
Family Structure (Two-parent, one-parent, stepfamily, etc.)	4.2	4	.38
Type of Referral (Status, property, person, drug, other)	3.71	8	.88
Socio-economic status	14.38	6	.03*
Variable	**F-RATIO**	**DF**	**PROB**
Age	.35	2/392	.79
Grade	.45	2/392	.64
+SRD School	.08	2/392	.92
SRD Property	.40	2/392	.66
SRD Drug	1.30	2/392	.27
SRD Person	.31	2/392	.74
SRD Total	.46	2/392	.63

+ Self-report delinquency

(the latter two did not differ significantly) at the one year follow-up at F $(2, 395) = 40.13$, $p < .05$. Diversion with services had a 22% recidivism rate; diversion without services a 32% recidivism rate; and traditional court processing a 34% recidivism rate. ADP resulted in a reduction in further official reports of delinquent acts.

THE THEORIZED PROCESS OF PROGRAM EFFECTS

The sequential process of labeling was addressed in one integrative path model where the relationship between diversion, perceived labeling from others, perceived expectancy of recidivism from others, perceived negative reputation, self-labeling, and delinquent behavior was assessed. Previous delinquency was expected to be related to labeling and later delinquency. It was expected that diverting youth from the juvenile justice process would result in less awareness of delinquency and labeling. The labeling variables were expected to be positively related to each other and to later delinquency. Stronger family relationships were expected to be related to less delinquency. A direct path from Family Relationships to delinquency was included to test which was stronger, the direct or indirect effect upon delinquency via labeling (Figure 1). The variables also represented data at multiple time points with some variables preceding others temporally. Coupled with the manipulation of group (Diversion) this strengthened the ability to make causal inferences though some limitations still existed. Variables collected at the pre-program time point were designated with the subscript "1," variables collected during the program with a "2" and post-program variables collected one-year after the pre-assessment with a "3." The labeling and family process variables were examined in relationship to delinquency during and after program participation.

THE REVISED EMPIRICAL PATH MODEL

The path models were analyzed using *EzPATH*, a computer program developed for *Systat* software (Steiger, 1989). Pearson correlation matrices were computed and then corrected for attenuation using a routine in the *Path* program by Hunter (1986). Table 3 presents both the raw and corrected correlations. The path model indicated that there was a

FIGURE 1. The Theoretical Model

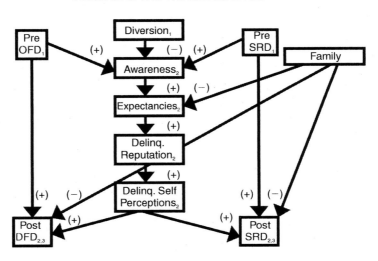

TABLE 3. Raw Input Correlation Matrices

	PRE OFD	DIVERT	GROUP	FAMILY RELAT	PERC AWARE	PERC EXPECT	PERC REP	SELF PERC	PST OFD
PRE OFD	1.000								
DIVERT	−0.487	1.000							
GROUP	−0.290	−0.023	1.000						
FAMILY RELAT	0.004	−0.018	0.055	1.000					
PERC AWARE	0.024	−0.033	0.057	0.022	1.000				
PERC EXPECT	−0.011	0.007	0.055	−0.194	0.262	1.000			
PERC REP	0.031	0.045	−0.066	−0.170	0.044	0.296	1.000		
SELF PERC	−0.001	0.028	−0.072	−0.090	0.021	0.213	0.520	1.000	
PST OFD	0.204	−0.013	−0.116	−0.017	−0.030	0.079	0.174	0.106	1.000

Key: PRE OFD = Pre official delinquency (court petitions) ; DIVERT = Level of Diversion (lo=court, med=intervention, high=warned/released); GROUP = Coded to examine program effects (lo =court condition, med= warned and released, high = intervention group); FAMILY RELAT = Family Relationships; PERC AWARE = Youth's perceptions of other's awareness of their delinquency; PERC REP = Youth's perceptions of other's labeling them as delinquent; SELF PERC = Youth's perceptions of self as delinquent; PST OFD = Post official delinquency (court petitions) good "fit" with the data, Chi-Square of 36.77, *df* = 18, ns. The final model is presented in Figure 2.

Diversion did not impact any of the labeling variables. Initial path coefficients from diversion to labeling ranged from −.03 to .03 and this variable was deleted from the model. Other program processes were found to be more influential than diversion alone. The amount of prior official delinquency did not seem to impact the labeling process though it was related to later official and self-report delinquency. Self-report delinquency was related to youth reporting that others were aware of and expected future delinquency (path coefficients of .17 and .26 respectively).

Family relationships were related to labeling and delinquency. Better family relationships were related to fewer perceived delinquent expectancies (−.35) and indirectly to less official delinquency. The direct path from family relationships to delinquency (coefficient of .04) was small and therefore deleted. Also, family relationships had a direct, inverse relationship to self-report delinquency (−.18). Youth who reported more positive family experiences reported less expected delinquency, less self-reported delinquency and indirectly less official delinquency.

The labeling variables were significantly related to each other and the theorized sequence of the variables was supported empirically. Greater perceived awareness of delinquency was related to higher perceived ex-

FIGURE 2. The Path Analytic Model

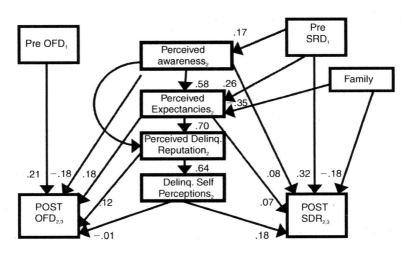

pectations for delinquency, more perceived delinquent perceptions and greater delinquent self-perceptions. The direct path coefficients between these four variables ranged from .58 to .70.

One interesting finding concerned the perceived awareness of delinquency variable. The direct path coefficient between Perceived Awareness and Perceived Negative Reputation was a negative value, $-.35$, while the indirect coefficient of Perceived Awareness through Perceived Expectancies to Perceived Negative Reputation was a positive value, .41. Thus, perceived awareness, others knowing about the youth's delinquent activity but not expecting more delinquent activity, was associated with youth feeling more positively about how others perceived them. This result likely reflects the positive effect of proactive awareness, without negative expectancies, in the lives of young people.

When examining the theorized pathways upon delinquency, interesting results emerged. First, the three variables measuring perceptions of other's attitudes had larger path coefficients to official delinquency than the youth's self-perception, which had a coefficient of $-.01$. This counters the possibility that these effects were solely a result of youth-reported data. If this were the case one might expect youth self-perceptions to outweigh youth's report of others' perceptions, in terms of relationships to delinquency. Further, perceived awareness of delinquency and perceived expectancies of future delinquency had different relationships to delinquency. Perceived awareness was directly related to less official delinquency $(-.18)$, while indirectly (through negative expectancies) was related to increased official delinquency (.18). Again, being aware of youth's delinquent activities was associated with decreased delinquency while awareness combined with negative expectancies was associated with higher levels of juvenile delinquent behavior.

DISCUSSION AND CONCLUSION

The findings in Phase 4 demonstrated the generalizability of the ADP model from a mid-sized, predominantly white small city to a larger, African American urban area. Phase 4 also demonstrated ADP's effectiveness with the use of paid staff from an existing agency, another potentially disseminable model, in addition to the previously validated use of volunteers. Importantly, the results indicated that there were no differences in recidivism between the warned and released group (diversion without services) and the traditionally court-processed or control group. Both resulted in more recidivism than family-based diversion. Based upon these findings, attending to the youth in the ecological context of their family and community is a more powerful approach to decreasing juvenile delinquency than either ignoring the act (diversion without services) or punishing the act via the juvenile justice system.

Labeling theory would posit that more diversion would result in less labeling and delinquency because of less exposure to labeling agents in the juvenile justice system. A substantial relationship to labeling and diversion was not detected. It could be argued that diversion is not the most salient variable. However, when exposure to labeling is assessed, labeling is more predictive of further delinquent acts.

The diversion program utilized here incorporated a family-oriented approach. Positive family relationships were related to reduced delinquent labeling and official delinquency. Further, youth who reported more positive family relationships also self-reported less involvement in delinquent activities. Of special significance in the current study was the differential impact of two of the labeling variables, perceived awareness of delinquency and perceived expectancies for future delinquency. While perceived awareness of youths' activities was related to *decreased* negative perceptions and behavior, perceived expectations for future delinquency was related to *increased* levels of both. Future tests of the conceptual model would be bolstered by including reports of labeling from multiple sources, in addition to the youth, to understand the process of labeling and its impact upon youth behavior. At this point, there seems to be no benefit to families, teachers, and friends stigmatizing young people by expecting delinquent behavior and communicating these expectations to young people (Cooper & Good, 1983). However, the findings supporting the positive function of awareness could be the result of the intentional and strategic incorporation of increased family monitoring and positive reinforcement of the youth's behavior into this behaviorally-oriented intervention (Bank, Patterson, & Reid, 1987; Patterson & Stouthamer-Loeber, 1984). The most extensive metanalysis to date demonstrates that behavioral family intervention programs have demonstrated substantial effectiveness in improving family relationships and decreasing delinquency (Lipsey, 1992). Parental monitoring and encouragement, without stigmatization, helps to reduce adolescent delinquency.

ADP seeks to reduce violence and delinquency by working with individual youth and the multiple contexts in their lives. The ecological programming model uses behavioral contracting with the youth and family.

Advocacy activities build upon youths' strengths and link to community assets. Volunteers and paraprofessionals are other resources available in most communities. The use of well-trained and supervised college service-learning students has facilitated the persistence of ADP. With dissipation of federal grant money for diversion programs, local and county governments have provided funding for this alternative based upon the use of volunteers. Should local agencies have staff and funding to implement the model, the current replication has found the use of paid staff to be efficacious as well. Setting is an important factor in that the program demonstrates effectiveness when located away from the juvenile justice context. In the work examining program content, incorporating behavioral contracting and community advocacy has been found to be more powerful than just attention from a well-intentioned volunteer with minimal training and supervision. This information on staffing, important components, and context has important implications for practice and replication.

Research has already demonstrated that program youth exhibit less delinquency a year later than those warned and released or those officially processed by the juvenile justice system. Further research could examine how long these results are sustainable with the current 3-4 month intervention approach. The more serious and violent offenders often have an earlier history of aggression and juvenile offenses (Loeber et al., 1993; U.S. Department of Health and Human Services, 2001). Future research could explore effects specifically upon violent behavior in adolescence and adulthood.

The effectiveness of ADP has been exhibited in both small town and large urban settings, mostly with white and African American youth. Further replication could examine other ethnic groups. Establishing the effectiveness of the model across families of diverse social and racial-ethnic origins is important in practice and dissemination efforts. In prevention science, it is important to demonstrate effectiveness with multiple groups and in diverse social, regional, and geographic settings.

In summary, ADP is a model based upon influencing the individual, family, community, and juvenile justice contexts. In randomized experimental studies, ADP has been shown to be a valid and effective alternative to juvenile court processing. This model has led to the actualization of one of the primary goals of ecological-community psychology: understanding a social problem and mobilizing a positive individual, family, and community response.

REFERENCES

Ageton, S.S., & Elliott, D. S. (1974). The effects of label processing on delinquent orientation. *Social Problems, 22*, 87-100.

Agnew, R. (1985). A revised strain theory of delinquency. *Social Forces, 64*, 151-167.

Agnew, R. (1993). Why do they do it? An examination of the intervening mechanisms between "social control" variables and delinquency. *Journal of Research in Crime & Delinquency, 30*(3), 245-266.

Akers, R. L. (1990). Rational choice, deterrence, and social learning theory in criminology: The path not taken. *The Journal of Criminal Law & Criminology, 81*(3), 653-676.

Alexander, J. F., Robbins, M. S., & Sexton, T. L. (2000). Family-based interventions with older, at-risk youth: From promise to proof to practice. *The Journal of Primary Prevention, 21(2)*, 185-205.

Bank, L., Patterson, G. R., & Reid, J. B. (1987). Delinquency prevention through training parents in family management. *Behavior Analyst, 10*(1), 75-82.

Berman, J.S., & Norton, N.C. (1985). Does professional training make a therapist more effective? *Psychological Bulletin, 98*, 401-407.

Cooper, H.M., & Good, T.L. (1983). *Pygmalion Grows Up: Studies in the Expectation Communication Process*. New York: Longman.

Davidson, W. S., & Johnson, C. (1987). *Diversion in Michigan*. Lansing, MI: Department of Social Services, Office of Children & Youth Services.

Davidson, W. S., & Rapp, C. A. (1976). Child advocacy in the justice system. *Social Work, 21*(3), 225-232.

Davidson, W. S., Redner, R., Amdur, R., & Mitchelle, C. M. (1988). *Alternative Treatments for Troubled Youth*. New York: Plunum.

Davidson, W. S., Redner, R., Blakely, C. H., Mitchell, C.M., & Emshoff, J. G. (1987). Diversion of juvenile offenders: An experimental comparison. *Journal of Consulting and Clinical Psychology, 55*, 68-75.

Davidson, W. S., Seidman, E., Rappaport, J., Berck, P. L., Rapp, N. A., Rhodes, W., & Herring, J. (1977). Diversion program for juvenile offenders. *Social Work Research and Abstracts, 13*(2), 40-49.

Downs, W. R., & Rose, S. R. (1991). The relationship of adolescent peer groups to the incidence of psychosocial problems. *Adolescence, 26* (102), 473-492.

Durlack, J. A. (1979). Comparative effectiveness of paraprofessional & professional helpers. *Psychological Bulletin, 86*, 80-92.

Eddy, J. M., & Gribskov, L. S. (1998) Juvenile justice and delinquency prevention in the United States: The influence of theories and traditions on policies and practices. *Delinquent Violent Youth: Theory and Interventions*. Thousand Oaks, CA: Sage.

Elliott, D. S., Huizinga, D., & Ageton, S. S. (1985). *Explaining delinquency and drug abuse*. Beverly Hills, CA: Sage.

Emshoff, J. G., & Blakely, C. H. (1983). The diversion of youth: Family-focused intervention. *Children & Youth Services Review, 5*(4), 343-356.

Gensheimer, L. K., Mayer, J. P., Gottschalk, R., & Davidson, W. S. (1987). Diverting youth from the juvenile justice system: A meta-analysis of intervention efficacy. In S. J. Apter & A. P. Goldstein (Eds.), *Youth violence* (39-57). New York: Pergamon.

Gray-Ray, P., & Ray, M. C. (1990). Juvenile delinquency in the Black community. *Youth and Society, 22*(1), 67-84.

Greenberg, D. F. (1999). The weak strength of social control theory. *Crime and Delinquency, 45*(1), 66-81.

Henggeler, S. W., Schoenwald, S. K., Borduin, C. M., Rowland, M. D., & Cunningham, P. (1998). *Multisystemic treatment of antisocial behavior in children and adolescents.* New York: Guilford.

Hirschi, T. (1969). *Causes of Delinquency.* Los Angeles: University of California Press.

Kantrowitz, R. E. (1979). *Training nonprofessionals to work with delinquent: Differential impact of varying training/supervision/intervention strategies.* Unpublished doctoral dissertation, Michigan State University.

Kaplan, H.B., & Johnson, R. J. (1991). Negative social sanctions & juvenile delinquency: Effects of labeling in a model of delinquent behavior. *Social Science Quarterly, 72*(1) 98-122.

Klein, M. W. (1986). Labeling theory and delinquency policy. *Criminal Justice and Behavior, 13*(1), 47-79.

Lipsey, M. W. (1992). Juvenile delinquency treatment: A meta-analytic inquiry into the variability of effects. In T. D. Cook, H. Cooper, D. S. Cordray, H. Hartman, L.V. Hedges, R. J. Light, T. A. Louis, & F. Mosteller (Eds.), *Meta-analysis for explanation: A casebook* (83-127). New York: Russell Sage Foundation.

Lipsey, M. W., & Wilson, D. B. (1998). Effective intervention for serious juvenile offenders: A synthesis of research. In Rolf Loeber and David Farrington (Ed), *Serious & violent juvenile offenders: Risk factors and successful interventions*, pp. 313-345.

Loeber, R., Wung, P., Keenan, I., Giroux, B., Stouthamer-Loeber, M., & Van Kammen, W. B. (1993). Developmental pathways in disruptive child behavior. *Development and Psychopathology, 5*, 101-132.

Matsueda, R. L. (1992). Reflected appraisals, parental labeling, and delinquency: Specifying a symbolic interactionist theory. *American Journal of Sociology, 97*(6), 1577-1611.

Mendel, R. A. (2000). *Less hype, more help: Reducing juvenile crime, What works and what doesn't.* Washington, D.C.: American Youth Policy Forum.

Patterson, G. R., & Stouthamer-Loeber, M. (1984). The correlation of family management practices and delinquency. *Child Development, 55*, 1299-1330.

President's Commission on Law Enforcement and the Administration of Justice. (1967). *Task Force report: Juvenile Delinquency and Youth Crime.* Washington, DC: U. S. Government Printing.

Schur, E. (1973). *Radical non-intervention: Rethinking the delinquency problem.* Englewood Cliffs, NJ: Prentice-Hall.

Smith, E.P., & Meyers, F. C. (1994). *Diversion of Youth from the Court System.* Grant proposal submitted to the Department of Public Safety-Juvenile Justice, funded for $217,500, October 1994-September 1997.

U.S. Department of Health and Human Services (2001). *Youth Violence: A Report of the Surgeon General.* Rockville, MD: U.S. Department of Health and Human Services, Centers for Disease Control and Prevention, National Center for Injury Prevention and Control; Substance Abused and Mental Health Services Administration, Center for Mental Health Services; and National Institutes of Health, National Institute of Mental Health.

Winfree, L. T., Backstrom, T. V., & Mays, G. L. (1994). Social learning theory, self-reported delinquency, and youth gangs. *Youth & Society, 26*(2), 147-177. August 20, 2002.

A Multidimensional Ecological Examination of a Youth Development Program for Military Dependent Youth

Daniel F. Perkins

Pennsylvania State University

Lynne M. Borden

University of Arizona

SUMMARY. The purpose of this article is to conduct a multi-method evaluation to investigate the effectiveness of a youth development program for military dependent youth entitled, Youth Action Program. The dimensions of an ecological program are used to examine the program post-de-facto. The comparison with elements of ecological program-

Daniel F. Perkins, PhD, is Associate Professor, Family and Youth Resiliency and Policy, Department of Agriculture and Extension Education, 323 Agricultural Administration Building, The Pennsylvania State University, University Park, PA 16802-2601 (E-mail: dfp102@psu.edu).

Lynne M. Borden, PhD, is Extension Specialist, and Associate Professor, School of Family Studies and Consumer Sciences, University of Arizona, P.O. Box 210033, Tucson, AZ 85721-0033 (E-mail: bordenl@ag.arizona.edu).

Address correspondence to Daniel F. Perkins.

This study was funded by a grant from the Virginia Polytechnic Institute and State University on behalf of the United States Air Force Family Advocacy Program (USAF FAP) and supported by The Penn State Agricultural Experimentation Project Number 3826.

[Haworth co-indexing entry note]: "A Multidimensional Ecological Examination of a Youth Development Program for Military Dependent Youth." Perkins, Daniel F., and Lynne M. Borden. Co-published simultaneously in *Journal of Prevention & Intervention in the Community* (The Haworth Press, Inc.) Vol. 27, No. 2, 2004, pp. 49-65; and: *Understanding Ecological Programming: Merging Theory, Research, and Practice* (ed: Susan Scherffius Jakes, and Craig C. Brookins) The Haworth Press, Inc., 2004, pp. 49-65. Single or multiple copies of this article are available for a fee from The Haworth Document Delivery Service [1-800-HAWORTH, 9:00 a.m. - 5:00 p.m. (EST). E-mail address: docdelivery@haworthpress.com].

Digital Object Identifier: 10.1300/J005v27n02_04

ming provided reasons as to this program's perceived impact and its weaknesses. A multi-dimensional evaluation is employed that examines the processes and outcomes in their natural settings. Youth and parents overwhelmingly believed that this program had made a positive difference in their lives. Youth demonstrated an improvement, although not significant, in their self-concept; however, participating youths' levels of social isolation and withdrawal from social contact remained at a high level. The results and need for future research are presented related to evaluating youth development programs and the ecological model. *[Article copies available for a fee from The Haworth Document Delivery Service: 1-800-HAWORTH. E-mail address: <docdelivery@haworthpress.com> Website: <http://www.HaworthPress.com> © 2004 by The Haworth Press, Inc. All rights reserved.]*

KEYWORDS. Evaluation, youth at-risk, prevention, military, ecological program

The United States military has over 1.3 million personnel, with more than half (55%) married (Mancini & Archambault, 2000). In addition, over 1.2 million children and youth from birth to age 18 live in military families (Mancini & Archambault, 2000), with approximately 300,000 of them being between 12 and 18.

In 1997, the Department of Defense surveyed 7,000 military teens and found that they had moved an average of five times (Kozaryn, 2000). The frequent moves, being separated from a parent on deployment, and the lack of long-term friendships present unique challenges to youth dependents of military personnel. These challenges place military youth dependents at greater risk for engaging in risk behaviors than their civilian counterparts, and thus they are in greater need of support and services. To address these issues, the Department of Defense provides supports and services through its youth development programs, such as the Youth Action Program (YAP) and recreation clubs.

Within the various purpose categories of ecological programs outlined in the article by Jakes (this volume), military youth development programs are implemented for multiple purposes: preventative, developmental, and individual empowerment. These programs offer young people one context for positive development, that is, an opportunity to engage in activities that meet their developmental needs (a developmental purpose) while decreasing the likelihood that they will engage in risky behaviors (preventative purpose; Benson et al., 1998). For example, youth who are involved have the opportunity to develop positive relationships that connect them to peers and adults in their communities. In a recent review of the literature, Scales and Leffert (1999) suggest that a positive relationship with an adult can be indirectly or directly related to higher levels of self-esteem and self-efficacy, reduced drug and alcohol use, and positive and improved school outcomes. Moreover, these contexts enhance the likelihood that youth will develop sustained and positive peer relationships, which, in turn, contribute to increased academic achievement (McLaughlin, 2000), increased social maturity, and buffered depressive symptoms (Scales & Leffert, 1999). Positive adult and peer relationships developed in conjunction with the structured activities provided by youth development programs increase the likelihood that youth will successfully navigate the challenges they face as they move toward adulthood (i.e., individual empowerment). McLaughlin's (2000) longitudinal study of youth and youth development programs provides strong evidence of the positive influence that participation in youth development programs can have on young people.

YOUTH DEVELOPMENT PROGRAMS AND ECOLOGICAL PROGRAMMING

The theories of change underpinning the military's youth development programs, or youth development programs in general, have often not been clearly articulated. Nonetheless, utilizing the words "youth development" with the word "program" implies common core concepts across programs. Of course, some programs adhere to those concepts better than other programs. The core concepts are found in the definition of community youth development. Community youth development is a process by which youth's developmental needs are met, their engagement in problem behaviors is prevented, and, most importantly, youth are empowered to build the competencies and skills necessary to become healthy contributing citizens now and as adults (Perkins, Borden, & Villarruel, 2001). Thus, by their very nature, effective and successful youth development programs are ecological programs. Yet, just as there is variance in youth development programs in their adherence to the core concepts of community youth development, there is also variance in how "ecological" these programs really are.

The purpose of this article is to examine a youth development program entitled Youth Action Program (YAP). Specifically, we conducted a multi-method evaluation to investigate the effectiveness of

YAP at reaching its goals. In this article, we first describe YAP. Then a brief examination of YAP in terms of the dimensions of an ecological program as outlined by Jakes (2004, this volume) is presented. The evaluation methods used in this study are presented next, followed by the results. Finally, we discuss the need for further research and the results in terms of evaluating youth programs and the ecological model.

PROGRAM MODEL

The Youth Action Program (YAP) was initiated in 1994 for the Department of Defense initiative *Model Community Project* through the collaborative efforts of the United States Air Force Academy and the Front Range Institute. (The Front Range Institute is a for-profit therapeutic center specializing in military families; however, the Institute's involvement in YAP was established under its not-for-profit corporation.) The purpose of YAP was to enhance academic performance, improve social skills, and reduce risk taking behaviors of military dependent youth participants (mostly 11 to 12 year olds). The ultimate outcome of YAP from the military's perspective included improved family adaptability and coping mechanisms, qualities assumed to be positively correlated with military career success. Potential participants were military youth dependents identified by school counselors and teachers as being "at-risk," a designation that involved such markers as failing grades or a recent drop in school performance, poor social skills development, behavioral difficulties, and/or minimal parental involvement. YAP attempted to improve family functioning and the young person's coping ability through mentoring, peer counseling, academic, and life skills training, and engagement in positive risk-taking experiences. Further, the Front Range Institute staff counselors and psychologists through YAP offered individual support and counseling for parents and also served as family advocates in school/family relationships. YAP's design emulated other successful programs in its attention to individual/youth life skills and its linkages to family and community services (Dryfoos, 1990, 1998; Lerner, 2002; Galavotti, Saltzman, Suter, & Sumartojo, 1997; National Research Council and Institute of Medicine, 2002; Perkins & Borden, 2003; Roth, Brooks-Gunn, Murray, & Foster, 1998).

YAP and Dimensions of the Ecological Model

In comparing YAP with the dimensions of the ecological model outlined by Jakes (2004, this volume), it is critical to remember that YAP was designed to increase the mission readiness of the military personnel. This is the coherent focus of all such military service programs as their youth development programs. The military reasoning is simply that a soldier is unable to be totally prepared for combat if he or she is concerned about a problem occurring within his or her family. When measured on its adherence to an ecological model, YAP seems to achieve high marks on seven of the twelve dimensions (i.e., program audience, program implementation, target of change, process of change, program purpose, program strategy, and program integration), while falling short on other dimensions (i.e., community control–program funding, community control–program development, community control–implementation, adaptiveness of program, and program institutionalization).

In terms of program audience, YAP scores high because it addresses three program audiences (i.e., youth, parents, and schools). For program implementation, YAP creates something new, but seems to fuse with the school on a limited, albeit important, basis. YAP's targets of change include the individual (i.e., youth and parent), microsystem (i.e, family), and the mesosystem (parent/child-school relationship). YAP's process of change focuses on changing a group of people to improve person/environment fit through incremental social change. As noted earlier, YAP has multiple purposes, including: preventative, developmental, and empowerment. YAP uses three methods, in that it works directly with youth, the parents, and the school. Finally, YAP demonstrates strong program integration, as the components of the YAP program are combined in a coherent, integrated fashion.

Where YAP seems to fall short in terms of the ecological model is in the community control dimensions. For example, YAP is at the consultation level for program funding and program development, that is, YAP attempts to meet the United States Air Force needs; however, there is no stakeholder and community representation, partnership, or citizen control. YAP has no participation from any outside agencies or persons in terms of program implementation. In terms of adaptiveness, YAP collects feedback through evaluation data, albeit inconsistently. However, YAP did not use that information to make changes to the program. YAP lacks program institutionalization, as demonstrated in the fact that it was terminated due to lack of continued funding.

YAP strategies focus on the formation and continuation of stimulating, supportive, and responsive environments; therefore, the program fits into the ecological programming paradigm. This comparison of YAP to the ecological dimensions will be utilized in the discussion section to provide insights about the results of the evaluation. Given that

YAP possesses many of the dimensions of an ecological program, a multi-method evaluation is desired in order to examine the dynamic nature of the program at the multiple levels of the ecology.

METHOD

The evaluation of the Youth Action Program (YAP) is composed of three major components: (1) process evaluation, (2) outcome evaluation, and (3) economic evaluation. This article presents the findings from the first two components. The process component included: development of a logic model, comparative analysis of best practices in youth programming, and focus groups and interviews with stakeholders. The first two parts (i.e., logic model and comparative analysis) of the process evaluation examined the program's change theory, the program elements, and the risk assessment process of YAP. The focus groups and interviews provided qualitative information from major stakeholders about the perceptions of YAP and its effectiveness.

The outcome evaluation examined what kinds of changes, if any, occurred in the participants' behavior. The outcome evaluation involved a follow-up telephone survey with past participants and a secondary data analysis. Originally, the outcome evaluation was also to include a match comparison longitudinal study; however, the data needed for the comparison group was not provided by the schools.

The implementation of this multi-method evaluation allowed for an examination of the processes and outcomes in their natural settings. Taken together, the data provide a multidimensional understanding of the effectiveness of the program in fulfilling its mission and meeting its measurable outcomes.

RESULTS

Process Evaluation

Logic Model

The logic model for YAP was completed through an iterative process (six exchanges) that involved the evaluation team and the YAP staff. In completing the process of developing YAP's logic model, one concern was identified. The YAP program did not appear to have a clear set of criteria for identifying potential participants. Generally, the program received its potential participants from the area schools for a variety of reasons, which included a number of key risk factors. Yet, sometimes the school officials were "concerned" about the individual without any concrete evidence to justify their concerns (e.g., poor grades, documented discipline problems, and/or documented social/emotional problems).

Comparative Analysis

A comparative analysis was conducted to verify and assess the major program elements of YAP. The comparative analysis was employed to determine the fit of YAP program components with common elements of successful youth prevention programs as identified by a comprehensive review of the research literature. The elements from this review were then adapted into a program elements checklist by the evaluation team. The checklist had nine categories that included: (a) individual/youth–life skills (11 elements); (b) individual/youth–future awareness and careers (3 elements); (c) family (3 elements); (d) community (3 elements); (e) collaboration (3 elements); (f) service delivery–philosophy/model (19 elements); (g) service delivery–program structure (11 elements); (h) service delivery–environment (11 elements); and (i) service delivery: staff development (4 elements). As noted in the parentheses, each of these categories had a range of program elements from three to nineteen. Respondents were asked to think about the degree to which a specific element was addressed by YAP. The range of responses was 1 = completely, 2 = mostly, 3 = somewhat, 4 = a little, to 5 = not at all. The scores were reversed in the analysis so that a high score indicated a high degree of integration.

Each member of the YAP staff (n = 4) and the evaluation team (n = 3) completed the checklist. Although a natural bias existed by having the YAP staff complete the checklist, the evaluation team thought it was critical to employ them in this assessment as a strategy to gain some insights from them, as well as to provide them with an opportunity to critically examine their program. Both the YAP staff and the evaluation team perceived a high degree of fit between the key elements of successful youth programs and YAP in three categories (a total of 11 out of 19 elements) of the program elements checklist. These included: (a) individual/youth–life skills: engagement of participants in intensive social skills training; opportunities for participants to develop problem solving skills, team work skills, a sense of mastery, goal-setting, and achievement; participation in positive risk-taking activities; doing well in school and grade achievement was stressed and supported by the program; (b) family: encouragement of family involvement in the program, removes obstacles or barriers to family involvement, establishment of positive support networks for families of participants; (c) service delivery-philosophy/model: integration into its program of a long-term preventative orientation, a clear mission, and a continuous evolution over time; designed to reach children early; established a flexible structure while maintaining high expectations of its participants; designed not to stigmatize participants.

The YAP staff and the evaluation team perceived a lack of fit between the key elements of successful youth programs and YAP in four categories from the program elements checklist. The elements included: (a) individual/youth–life skills: lack of opportunities for participants to engage in service activities; lack of opportunities for youth to be involved in the planning, carrying-out, and assessment of the program; (b) individual/youth–future awareness and careers: lack of opportunities for youth to explore career options; lack of encouragement of youth's curiosity about work and earning money; (c) collaboration: lack of involvement of police officers and "street-wise people" in programming; lack of media resources for publicity such as public service announcements, recruitment, and acknowledgments; (d) service delivery–philosophy/model: limited access for participants to resources such as career planning, sex education. The lack of fit of these elements distinguishes areas that YAP needs to address for program improvement.

Focus Groups

Focus groups were conducted with families of participants. Using the same questions as in the focus groups, individual interviews were conducted with the following stakeholders: parents of program dropouts or non-joiners, mentors, school personnel, and military personnel who had subordinates in the program. However, only the results of the focus groups and interviews with parents of program dropouts or non-joiners are presented here, due to space limitations. Patton (1987) suggests that, ideally, focus groups should have six to eight members who form a relatively homogeneous group (p. 135). For our purposes, a random sample of 25 parents from the families (N = 211) whose children (N = 235) had completed the program were invited, via a letter sent to them by YAP facilitators, to participate in a two-hour focus group session designed to elicit their perspectives on the program's effectiveness in producing changes in their children. The decision to include 25 families was based on budget considerations and the logistics of having only four days to complete the work in the field.

Out of the 25 invitations sent, nineteen families agreed to have at least one member participate; the other six families had either moved out of the area or were unable to participate due to work and family schedules. YAP staff organized the participants into three groups, each representing five to seven families, and scheduled them into one of three consecutive evening groups at their convenience. Although the groups were homogeneous in that all members' children had completed either a one-year or an 18-month program (the difference being that the longer program served older cohorts), the parents included retired and active duty personnel, a single-parent female, and the spouses of retired and active duty personnel, and represented both noncommissioned and junior officers in the military hierarchy. We limited our focus groups to parents whose children had already completed either one year or eighteen months in the program, since we felt that as active stakeholders they would provide a broader, long-term perspective on family relationship issues and increase our understanding of those issues in regard to their military careers. Our goal in implementing focus groups was to gain insight and understanding into the participants' perceptions of the program experience in relation to the everyday life of military families; thus, we chose a less structured approach as our focus group strategy. The questions were designed to elicit an ongoing dynamic within the group rather than reach "findings of fact," as in traditional scientific research models. For example, we opened the focus group by asking parents what they thought was the reason for their child's referral to the program, with a follow-up question of why they, as parents, decided to commit to it.

In the interests of increasing analytical consistency, one member of the evaluation team, who specializes in narrative data analysis, conducted the three parent focus groups and used insights gained from the previous evening's discussion as stimulus for more in-depth discussion with the following evening's group. All focus group sessions were audio taped, and transcriptions were prepared from the tapes. Data from transcripts of the focus groups were analyzed with an emphasis on emergent themes and the dynamic interaction through which they were constructed.

Perceptions of Youth Outcomes. While not all parents reported grade improvement, they were unanimous in describing positive social and academic outcomes from their children's participation in YAP. Typically, they noted behavioral changes that continued even after the culmination of the program. They attributed these changes first to the team-building activities that focused on developing trust, and, secondly, to the emphasis placed on skill-building in the weekly group meetings. As one parent said: "It helped [my daughter] with her organizational skills tremendously, with the homework and getting it in on time . . . " Overall, increased social development ranked high with parents as a noticeable outcome of the program. One father, whose son was a good student before entering the program, noted that, although his son's organizational skills had not been improved by the program, YAP had provided important social links in the boy's life. Several parents expressed a view of YAP as helping their children find a place within the broader social context–a big connect. Parents frequently noted a child's increasing self-confidence among people, or their developing sense of self-worth and belonging. "I think she was a lot happier," a mother explained, after describing the difficulties her daughter had encountered when they first moved from the military base into the community.

This urgency of "a big connect" was demonstrated repeatedly as participants related their stories of family disruption and disconnection as they moved to new military posts or during periods when fathers or mothers were called away to active service. Finding connections in new schools and new communities proved to be a goal for each of these families in a uniquely military way.

Most parents also included issues of anger management and trust as areas that showed positive changes in their children. "He's not doing the explosive thing and he doesn't run out of the house and disappear," one parent said. Parents described how the bonds of trust that developed around their children's mutual understanding of one another's needs and the aura of confidentiality that supported those bonds were important to the progress their children had made.

Perceptions of Changing Family Life. Parents were unanimous in crediting the program with improving relationships between themselves and their children, as well as with improving the emotional climate of family life as a whole. Several parents who had originally described how the program came into their lives when they were "at the end of their rope" now spoke of the program as "the light at the end of the tunnel." When the facilitator asked what they had gained from the program, these parents readily described the ways in which their families had benefited. When parents talked about improved relationships with their children, for example, they frequently spoke in the context of program activities. Camping and wall climbing were the activities most

mentioned in these discussions. The conversations were usually ani-
mated and punctuated by a kind of insider laughter which appeared to
show relationships developed through shared experience.

Perceptions of YAP's Role in Relation to Military Life. Program evaluators also examined the supportive aspects of the program–how YAP
acted as a support group for families and how this support affected military careers. The active or retired military parents talked about their
concerns for spouses and children at home, while the support parents,
who handled the home and family activities during times of family sep-
aration, elaborated on issues of single-parenting. Whether active, retired, or a supporting spouse, all parents recalled their experiences as the
absent parent or the parent left behind to handle the day-to-day matters
of the family. Shifting the focus of the conversation from children's
fears to the emotional tensions of single parenting, one parent said:
"Don't assume that these [programs] are just for supporting kids. These
[programs] are also for the parent. It gives them another additional
mechanism to talk to somebody."

However, parents emphasized how the primary beneficiaries of the
program were the children, who, when parents were away on a tour of
duty, would have activities to anticipate and adventures to relate to the absent parents in letters, email correspondence, or during telephone conversations. "At least with a program like this they could see that, in two
weeks we're going to do this, in another three months we go to [the camping site], and another couple months we do the horseback riding, and then
we do this, and then dad's home again. . . . " In this way, they portrayed
YAP as a vehicle for establishing continuity in their children's lives. As a
coping mechanism, the scheduled events became an organizing principle
through which families regained a sense of personal and family order.

Interviews with Parents of Program Drop-Outs and Non-Participants

YAP staff from the Front Range Institute provided the names and
phone numbers of all the youth (i.e., 16 youth) and their parents who were
invited to join the Youth Action Program but chose not to participate (i.e.,
8 families) or who dropped out before completing the program (i.e.,
8 families). Several attempts were made to contact each of these families
at work or at home. Two parents refused the interview, four home numbers
were disconnected and work contact numbers were no longer accurate.
One family had a call blocking mechanism that screened out our call. In
all, we were able to speak directly to five parents (31% completion rate).
The low completion rate was expected for two reasons: (1) the inherent

mobility of military families and (2) the lack of attachment of this non-participatory population to the program. Children in three of the families had participated in at least one YAP group meeting.

The interviews with these parents were open-ended, that is, they focused on the question of why the family had chosen not to participate in YAP. Follow-up questions were introduced as needed to gain a better understanding of parents' perspectives. For those who had attended at least one session, the evaluators attempted to find out what originally had appealed to them about the program. Although the sample number is small, the perceptions of this group of non-participating or drop-out parents are meaningful when compared to the perceptions of participating parents. The five interviews completed provided a range of reasons for non-participation, as well as a widely varying range of perceptions about the program itself. Parents frequently reported more than one reason for non-participation or provided additional commentary on the program. Location in terms of convenience of driving from home or work to the YAP meeting place and scheduling around employment, school, church, and other family activities were concerns of two parents, although both included other reasons for non-participation as well. Other reasons that parents gave for not participating included: Program concentrated on the youth and not enough on the family, children in the program were perceived to be a negative influence, the child felt "out of place" and "dragged his heels" on attendance, and lack of communication (e.g., parent was appalled that the child refused to talk about group meeting discussions; YAP staff never explained why their child was selected).

OUTCOME EVALUATION

The outcome evaluation component consisted of the secondary data analysis and the telephone survey. Because the secondary data was (was or were) not the same assessments from year to year, and for organization and simplicity in reading, the secondary data analysis information was presented by cohort year. Data for cohort years 1997 and 1998 was not available. The telephone survey included both analyses of past YAP youth participants as well as past YAP parent participants.

Secondary Data Analysis

The Personality Inventory for Youth (PIY) was used to assess children's behavior, affect, cognitive status, and family characteristics. The

PIY uses empirically replicated correlates based on evaluations carried out by teachers, parents, and clinicians. This measure was administered to the 1996 cohort using a pre/post test design. At the start of the first YAP session, youth completed this instrument. Then at the last session, participating youth completed this same instrument. Data was available for 32 youth, of who the average age was 12.9 and 53% were male. To compare changes in individuals' scores from the pretest to the posttest, a mean was calculated for each child and a paired T-test analysis was performed on each scale. Three scales out of 13 had close to significant differences, with slight increases or decreases with participation in the program. All changes in the scales were in the positive direction in terms of desired outcomes. These scales were the Delinquency (decreased; p = .059), Social Withdrawal (decreased; p = .076) and Behavioral Inconsistency (decreased; p = .066) scales.

The Berks Behavior Rating Scale was used to measure children's behavioral problems. This measure was administered to the 1999 cohort using a pre/post test design. At the start of the first YAP session, youth completed this instrument. Then at the last session, participating youth completed this same instrument. Data was available for 23 youth, of who the average age was 14.2 and 48% were male. A comparison was made between the pretest and posttest results. This was achieved by creating cross tabulations for each scale, comparing the percentages of youth who were rated as having high scale scores with those who did not. The overall trend depicted a positive increase in incidence of appropriate behaviors for all scales from pretest to posttest. Increases greater than or equal to 10% were observed in 15 out of the 19 scales. The highest increases were observed in the Poor Ego Strength scale, with a change of 30% between pretest (32%) and posttest (62%) youth. The Poor Impulse Control scale exhibited a similar change. Thirty-four percent (34%) of pretest youth were identified as having high scale scores, and that increased to 60% for posttest youth. Thus, youth who participated in YAP reported gains in their self-confidence about their abilities and in their ability not to respond impulsively. The Poor Attention scale had the third highest percent change (24%), from 46% pretest youth reporting an ability to focus and be attentive for a period of time to 70% of the youth reporting this ability after participating in the program. The fourth largest increase occurred in the Excessive Dependency scale, where 16% of pretest youth were identified as having high scale scores, that is, they did not have an exaggerated need to gain support from others. After participating in the program, the number of youth who reported an appropriate need for support had increased to 38%.

Given the small sample size, caution must be applied in interpreting these results. Moreover, given that there was no control group, it is not possible to say for certain whether it was the program that caused these changes. However, it is important to note that the changes were probably not simply due to maturation, because similar changes were not evidenced when examining the normal ranges of these measures.

Telephone Survey

The follow-up telephone survey was conducted to assess YAP participants' engagement in positive activities, risk behaviors, and perceptions of their relationship with parent(s). Conducted in March of 2000, the telephone survey made use of a brief structured interview. The YAP staff provided a list of potential respondents' telephone numbers to the evaluation team. The evaluation team then made three attempts to contact potential respondents. If the number had changed, the evaluation team located the correct number whenever possible. The response rate for the telephone survey was 48% (87/180). Forty-two percent, or 75, of the past YAP participants declined to partake in the telephone surveys. When asked why they did not want to participate, the majority of them reported lack of time. Approximately eight potential respondents indicated their unwillingness to respond due to their extreme dislike of the program. These individuals noted that the program seemed to be fostering values that were against their family such as: secrecy, over dependence, and excessive emotionality. Another 10% of the YAP participants could not be located.

The items and scales from this survey were drawn from other national surveys, such as the National Youth Longitudinal Survey, Monitoring the Future, and the Adolescent Health Longitudinal Study. The scales were created by the average score that was determined by adding the values of the responses offered for each question that made up each scale. Missing values were excluded from analysis and were not computed in the average score. The MEAN function in SPSS was used to ensure that each scale was divided by the appropriate number of cases.

The scales for this survey included: Parent adolescent communication scale; satisfaction scale (students' satisfaction with their academic work), social alienation scale; truancy scale, deviant behavior scale, unprepared scale (students' preparedness to perform at school), future education plans, involvement in community activities (e.g., sports, clubs). For the majority of the scales, YAP participants reported within the normal range, that is, similar to other adolescents who completed the above-noted surveys. Their scores on one scale, the satisfaction scale, were slightly less than the national average for that scale. Thus, it appears that the past YAP participants are functioning within the normal range of most American teenagers. This analysis cannot address issues of change, as no pretest was administered.

Parents also answered several questions during the follow-up survey. Approximately half of the parents reported positive changes. For example, 47% of respondents indicated that their relationships with their children had improved since their participation in YAP. Fifty-two percent indicated that their child's schoolwork had improved. When asked whether their child's attitude towards school had changed since participating in YAP, 53% of the respondents indicated that it had changed for the better. When asked to describe how their child's attitude had changed since their participation in YAP, an overwhelming number of parents (90%) indicated that their child's behavior had changed in a positive manner. When parents were given the opportunity to express any comments or ideas they might have regarding the YAP program, 47% responded to the open-ended question, and a majority of those parents who responded (78%) expressed positive comments, while only 22% expressed negative comments. Yet, most of the parents (92%) indicated that their participation in YAP had no impact on their decision to stay in the Air Force.

DISCUSSION

The use of multiple evaluation methods afforded a more comprehensive examination of this youth program by examining the processes and outcomes in their natural settings. By examining the various levels of the youth's ecology related to this program, this evaluation was able to detect meaningful findings. For example, the telephone survey and the focus groups with youth and parents provided evidence that YAP had made a *perceived* positive difference in the lives of participating youth and families. The secondary data analysis provided mixed evidence of success in a few outcome areas. For example, youth improve, albeit not significantly, from pre- to post-program in their self-concept and in their self-control as measured by the PIY. Yet, there was no decrease in the youth's levels of social isolation, withdrawal from social contact, antisocial behavior, or expression of anger. The logic model and the comparative analysis uncovered real strengths within the program (e.g., trained staff, skill-based, and experiential learning methods) and also identified some of the program's weaknesses (e.g., no clear set of criteria for identifying potential participants, lack of opportunities for participants to engage in service activities, and lack of opportunities for youth to be involved in the planning, carrying-out, and assessment of the program). The mixed findings of this evaluation suggest that the program is working in some ways and not working in other ways. Clearly, parents and youth perceive that they and their families are better off because of YAP and that the program allows them to be mission-ready, but the outcome data does not provide the same picture. The outcome data suggests that the program may have some short-term influence on some behaviors, but not on many.

As argued earlier in this manuscript, youth development programs that are effective are by nature ecological programs. Of course there is variance in how "ecological" a program is, like YAP, with regard to the program elements. Clearly, the success of the program as measured by the perceptions of its closest audiences indicates that it was having a positive impact. YAP's focus on multiple audiences and its implementation of several strategies make it an excellent example of an ecologically based program. Moreover, YAP strategies address multiple levels of the ecology within the military family. Most importantly for YAP's perceived success by parents and youth, YAP's strategies focus on the formation and continuation of stimulating, supportive, and responsive environments. These reasons drawn from the comparison with elements of ecological programming provide some understanding of why YAP was able to have so great a perceived impact.

The weaknesses of YAP, especially with regard to the community control dimensions, provide an explanation for YAP's closure after extramural funding ended. The lack of community ownership in this program meant the demise of it in the end, and may be in part because it lacked grassroots advocacy or because YAP was not well known in the community nor did it draw in citizens into its program. Moreover, the lack of solid quantitative data available for this evaluation is due in large measure to the absence of YAP's adaptiveness. Although feedback was collected, it was inconsistent and it was not examined by YAP staff.

Although this investigation examined YAP from multiple levels of the youth's ecology and provided useful information, further research is necessary. First, YAP, and other youth programs, for that matter, may want to examine their programming in terms of the dimensions of ecological programming. In doing so, the program staff may be immediately strengthening their program and increasing its chances for success. Second, a longitudinal study that employs an experimental design is necessary to assess YAP's impacts beyond perceptions, as well as to understand which of YAP's processes are important to its effectiveness and which are not as influential. In addition, a longitudinal study is required to determine whether sleeper effects may exist (National Research Council and the Institute of Medicine, 2002). Within an ecological perspective, then, this type of evaluation research needs to pay special attention to the factors at the various levels that influence the effectiveness of the practices and components within YAP.

REFERENCES

Dryfoos, J. G. (1990). *Adolescents at risk: Prevalence and prevention.* New York: Oxford University Press.

Galavotti, C., Satlzman, L., Sauter, S., & Sumartojo, E. (1997). Behavioral Science Activities at the Centers for Disease Control and Prevention. *American Psychologist, 52(2),* 154-166.

Jakes, S. S. (2001). *Understanding ecological programming: Evaluating program structure through a comprehensive assessment tool.* Unpublished doctoral dissertation, North Carolina State University, Raleigh, NC.

Kozaryn, L. D. (2000). Department of Defense to assess youth support. American Forces Press Service, September 7. Retrieved December 21, 2000 from: http://www.defenselink.mil/news/Sep2000/n09072000_20009073.html.

Lerner, R. M. (2002). *Adolescence: Development, diversity, context, and application.* Upper Saddle River, NJ: Prentice Hall.

Mancini, D. L., & Archambault, C. (2000). What recent research tells us about military families and communities. Presentation at the Department of Defense's Family Readiness Conference. Phoenix, AZ.

McLaughlin, M. (2000). *Community Counts: How youth organizations matter for youth development.* Retrieved January 12, 2001 from http://www.PublicEducation.org. Washington, DC: Public Education Network.

National Research Council and Institute of Medicine (2002). *Community programs to promote youth development.* Committee on community-level programs for youth. Board on Children, Youth, and Families, Division of Behavioral and Social Sciences and Education. Washington, DC: National Academy Press.

Perkins, D. F., & Borden, L. M. (2003). Key elements of community youth development programs. In F. A. Villarruel, D. F. Perkins, L. M. Borden, & J. G. Keith (Eds.), *Community youth development: A framework to promote individual and collective well-being* (327-340). Thousand Oaks, CA: Sage.

Perkins, D. F., Borden, L. M., & Villarruel, F. A. (2001). Community youth development: A partnership for change. *School Community Journal, 11,* 39-56.

Roth, J., Brooks-Gunn, J., Murray, L., & Foster, W. (1998). Promoting healthy adolescents: Synthesis of youth development program evaluations. *Journal of Research on Adolescence, 8,* 423-459.

Scales, P., & Leffert, N. (1999). *Developmental assets: A synthesis of the scientific research on adolescent development.* Minneapolis, MN: Search Institute.

Cultivating Capacity:
Outcomes of a Statewide Support System
for Prevention Coalitions

Roger E. Mitchell
Brenda Stone-Wiggins

North Carolina State University

John F. Stevenson
Paul Florin

University of Rhode Island

SUMMARY. Although community coalitions are an increasingly popular mechanism for attempting to change community-wide health, the

Roger E. Mitchell, PhD, is Assistant Professor, Department of Psychology, Box 7801, North Carolina State University, Raleigh, NC 27695-7801 (E-mail: Roger_Mitchell@ncsu.edu).

Brenda Stone-Wiggins, MPH, is affiliated with the Department of Psychology, North Carolina State University, Raleigh, NC.

John F. Stevenson, PhD, Professor and Director, Doctoral Program in Experimental Psychology, Department of Psychology, University of Rhode Island, Kingston, RI 02881 (E-mail: jsteve@uri.edu).

Paul Florin, PhD, is Professor in Psychology, Department of Psychology, University of Rhode Island, Kingston, RI (E-mail: pflorin@uri.edu).

Address correspondence to: Roger E. Mitchell.

The ideas in this manuscript emerged from work that was supported in part by a grant from the Center for Substance Abuse Prevention. The authors would like to thank Kristine Chadwick and other members of the evaluation team for their dedication. Also thanks to anonymous reviewers for comments that allowed the authors to strengthen this manuscript.

[Haworth co-indexing entry note]: "Cultivating Capacity: Outcomes of a Statewide Support System for Prevention Coalitions." Mitchell et al. Co-published simultaneously in *Journal of Prevention & Intervention in the Community* (The Haworth Press, Inc.) Vol. 27, No. 2, 2004, pp. 67-87; and: *Understanding Ecological Programming: Merging Theory, Research, and Practice* (ed: Susan Scherffius Jakes, and Craig C. Brookins) The Haworth Press, Inc., 2004, pp. 67-87. Single or multiple copies of this article are available for a fee from The Haworth Document Delivery Service [1-800-HAWORTH, 9:00 a.m. - 5:00 p.m. (EST). E-mail address: docdelivery@haworthpress.com].

http://www.haworthpress.com/web/JPIC
Digital Object Identifier: 10.1300/J005v27n02_05

empirical evidence has been mixed at best. Technical Assistance (TA) efforts have emerged in greater scale in hopes of improving both programming quality as well as the coalition structures supporting such programs. However, this commitment to TA interventions has outstripped our knowledge of optimal ways to deliver such assistance, and its limitations. This study takes advantage of results from a state-wide technical assistance project that generated longitudinal data on 41 health-oriented coalitions. The following questions were addressed: What are the circumstances under which coalitions will utilize available assistance? What are the effects of technical assistance on intermediate community outcomes? The results suggested that coalitions with greater initial "capacity" used more TA. Coalitions with low utilization mentioned difficulty in identifying their TA needs as the salient reason for not pursuing these resources. Over time, there were significant positive changes in coalition effectiveness as perceived by key informants, but these were not influenced by amount of TA. *[Article copies available for a fee from The Haworth Document Delivery Service: 1-800-HAWORTH. E-mail address: <docdelivery@haworthpress.com> Website: <http://www.HaworthPress.com> © 2004 by The Haworth Press, Inc. All rights reserved.]*

KEYWORDS. Community coalitions, substance abuse prevention, Technical Assistance, community intervention, prevention

Ecological frameworks have long served as touchstones for those interested in prevention and health promotion. Ecological approaches focus attention on the reciprocal relationships between individuals and contexts, and compel one to look at multiple levels of influence in understanding the etiology of health problems (McLeroy, Bibeau, Steckler & Glanz, 1988; Stokols, 1992; Winett, 1995). From this perspective, preventive interventions are more likely to be successful when there is an alignment of intervention strategies at the individual and environmental level (e.g., Dishion & Kavanagh, 2000). The notion of community level solutions has also become popular, for somewhat different reasons, among those concerned with empowerment, community capacity and, more recently, social capital. These concepts stress the importance of supporting grassroots participation, engaging multiple constituencies, increasing inter-organizational linkages and strengthening community problem-solving (e.g., Connell, Kubisch, Schorr, & Weiss, 1995; Minkler, 1997). Efforts to bring together a broad range of community constituencies in addressing health problems also appeal to our sense of participatory democracy (Green & Kreuter, 2002). Together, these rationales have stimulated public agencies and foundations to sponsor a variety of multi-component, community-level interventions over the past decade (Wandersman & Florin, in press).

Community-level health interventions often employ organizational structures such as community coalitions, partnerships or task forces to mobilize multiple community sectors to address cardiovascular risk, substance abuse, and other health outcomes. However, reviews and cross-site evaluations show a modest and mixed record of community-based coalition interventions achieving desired community-wide health outcomes (Roussos & Fawcett, 2000; Saxe et al., 1997; Sorensen, Emmons, Hunt, & Johnston, 1998; Merzel & D'Affitti, 2003; Wandersman & Florin, in press). A variety of reasons have been offered for such disappointing results, including methodological limitations of evaluations, inefficiency of coalitions as a change mechanism, the failure of coalitions to utilize and effectively implement "evidence-based" programs, and the tendency to gravitate towards strategies targeted at individuals rather than those targeted at institutional level policies and practices (Hallfors, Cho, Livert, & Kadushin, 2002; Kreuter, Lezin, & Young, 2000; McLeroy, Norton, Kegler, Burdine, & Sumayo, 2003).

One response to such failures has been to consider technical assistance and training (TA) that might improve both the programs implemented as well as the viability of the coalition structure supporting such programs (Fawcett et al., 1995; Florin, Mitchell, & Stevenson, 1993; Manger, Hawkins, Haggerty & Catalano, 1992; Mitchell, Florin, & Stevenson, 2002). With many possible reasons for variability in coalition effectiveness, TA is one strategy available to policy makers attempting to actualize the benefits of the coalition approach. Research on coalition structure and processes suggests a set of core competencies (e.g., member recruitment and leadership development, meeting facilitation, action planning, program development and implementation, evaluation, social marketing and fund-raising) that seem logically linked to coalition functioning (Butterfoss, Goodman, & Wandersman, 1996; Florin, Mitchell, & Stevenson, 2000; Kegler, 1998). For example, Hays, Hays, DeVille, and Mulhall (2000) found that the degree of sectoral representation in a coalition was related to the development of more effective plans. Similarly, Florin et al. (2000) found that organizational climate and coalition linkages predicted key informants' ratings one year later of coalition effects on community norms, policies and prevention resources.

The focus of much TA is to build capacity so that coalitions are in a better position to accept and implement science-based prevention. In the programming realm, use of evidence-based programming has become the standard against which coalition intervention efforts are increasingly judged. However, funders are acknowledging the challenge of adapting evidence based programs to local conditions and circumstances without destroying the "effective ingredients." Green and Kreuter (2002) state, "Many funding agencies have found a comfort zone between the rank empiricism of imposing best practices and the rank localism of ignoring previous research in favor of indigenous judgment . . . A common form is diplomatically offering and funding technical assistance in support of the funded communities" (p. 305). Increasing efforts are being made to build the capacity of coalitions to recognize, use and assess evidence-based programs (Gibbs, Napp, Jolly, Westover & Uhl, 2002; O'Donnell et al., 2000). Feinberg, Greenberg, Osgood, Anderson, and Babinski (2002), for example, describe presentation of a risk and protective factor framework as a means of increasing program impact. In a case study of on-site technical assistance to 13 community-based organizations (CBOs), Stevenson, Florin, Mills and Andrade (2002) found that utilizing greater levels of technical assistance was associated with greater evaluation capacity, which was intended to support more "evidence-based" interventions.

The belief in the potential efficacy of technical assistance and training has led governmental and nongovernmental organizations to create state-wide or regional structures that would more effectively support coalitions or other community-based organizations providing prevention services. Through a variety of grant and contract mechanisms, the Center for Substance Abuse Prevention (CSAP) has committed substantial resources to provide on-site and off-site technical assistance and training support to local prevention initiatives (e.g., Schinke, Brounstein & Gardner, 2002). One CSAP effort has involved the development of a web-accessible registry of prevention programs that is searchable by target population, type of intervention, and so on (CSAP, n.d.). The State of Pennsylvania has committed $1.5 million to train local communities in a risk and protective factor community change model (Feinberg et al., 2002). With a goal to strengthen HIV/AIDS prevention efforts among disproportionately affected minority communities throughout the United States, CDC has provided customized technical assistance to CBOs regarding evaluation capacity (Davis et al., 2000). In the non-governmental realm, the Community Anti-Drug Coalition of America (CADCA) is an intermediary membership organization of over 5,000 anti-drug coalitions that supports local efforts to make their communities healthier. CADCA offers web-based technical assistance packets with a list of national organizations, publications, community stories and community leaders with an expertise in 18 areas of interest. Francisco et al. (2001) describe development of a web-based "Community Tool Box" (CTB) designed as a resource to bolster "core competencies" of community building. The CTB received over 2 million "hits" in 1999.

This study takes advantage of a state-wide project in Maine that provided technical assistance to local coalitions. Longitudinal data was collected from multiple sources (i.e., coalition leaders, community key informants) on 41 health-oriented coalitions in order to address two questions. First, what are the circumstances under which coalitions will utilize available assistance? If use of technical assistance is voluntary, the coalition must see the utility of such assistance. Securing appropriate technical assistance may be tied to the coalition's understanding of its own strengths and weaknesses, which may not occur until later in its development. Are the most well-organized and effective coalitions the ones most likely to use available resources? In this study, we will examine the "penetration" of the TA intervention and whether greater levels of paid staff, coalition capacity, and community linkages are associated with greater use of TA in our sample of coalitions.

Second, what are the effects of technical assistance on intermediate outcomes? One of the complaints about the "coalition movement" is that it created unrealistically high expectations of the kind of effects coalitions could have. A closer examination of the intermediate effects of technical assistance (e.g., on coalition functioning and programming) might help us do a better job of estimating the amount of technical assistance needed to have an effect at different points in the process. We will examine whether the amount of technical assistance utilized is associated with changes over a 15 month period in key informants' rating of coalition effects in the community. We will also examine effects of varying technical assistance on the extent of increased linkages across coalitions.

In summary, this study will examine these specific questions: Do time 1 coalition characteristics (e.g., expressed need for technical assistance, hours of paid staff, length of existence, coalition capacity, and coalition contacts) influence utilization of technical assistance in the subsequent 15 months? Are amounts of technical assistance utilized associated with pre-post changes in key informants' rating of coalition effects?

METHOD

Project Background

The Dirigo Prevention Coalition originated as a 3-year federal demonstration grant funded by the Center for Substance Abuse Prevention to build the capacity of local coalitions. (The name of Dirigo is taken from the State of Maine motto, "I lead.") The original state agency partners in the grant proposal (i.e., Bureau of Health, Office of Substance Abuse Prevention, and Department of Corrections) were already funding a diverse array of local coalitions dealing with such issues as tobacco control, substance abuse, juvenile delinquency prevention, breast and cervical health, cardiovascular disease, and community health. In some communities, local coalitions had overlapping membership and missions, but minimal coordination. The objectives of the project were to provide training and technical assistance directly to community-based coalitions, to encourage greater collaboration among diverse coalitions at the local level, and to encourage greater cooperation among the state agencies funding these local coalitions.

Four project staff members with expertise in training, evaluation, and community development began making contact with target coalitions in 1996. They tried to engage a diverse range of coalitions through training sessions, regional forums, a website, newsletter, and individualized technical assistance by phone, by mail, and/or on site. One goal of all these efforts was to promote greater connections and networking at the local level among coalitions. At the level of state practices and policies, the staff advocated for increased resources for local efforts and for increased coordination among state agencies. The evaluation of these efforts continued through the end of 1998.

Coalition Sample

The original state agency partners in the grant (i.e., The Bureau of Health, the Office of Substance Abuse, and the Juvenile Justice Advisory Group) supplied the names of 52 coalitions funded by their offices. These coalitions dealt with such issues as smoking (ASSIST sites), substance abuse (OSA coalitions), juvenile delinquency prevention (JJAG coalitions), breast and cervical health (BCHP coalitions), cardiovascular disease (CVD coalitions), and community health (PATCH and Healthy Community coalitions). To this list, we added two additional coalitions identified through the course of our interviews. Contact with coalition coordinators indicated that 4 coalitions were inactive (i.e., had not met in the past 12 months) and that 5 coalitions were listed more than once, since they received funding from more than one of these state funding sources. Four groups were implementing prevention programming, but did not consider themselves to be coalitions (i.e., they were staff who were implementing grant supported prevention but without broad community input). This resulted in an initial sample of 41 coalitions with whom we conducted baseline leader interviews in 1996.

These initial 41 coalitions were spread throughout the state, covering 15 of the 16 counties in Maine. Approximately 37% had been in existence 1 to 2 years, 32% had been existence 3 to 5 years, and 31% in existence 6 or more years. In terms of geographic coverage, 18 served a single city (or municipality), 11 served more than one municipality, 7 served a county, and 5 served more than one county. This sample allowed us to look at issues of initial utilization.

Study Design

We considered several choices for a research design to assess technical assistance effectiveness. For reasons of cost and uniqueness, it would have been difficult to find another state with similar coalitions that could serve as a comparison site. An alternative was use of a "wait-list" control group of coalitions from within the state that would have provided a comparison group. However, project staff members felt strongly that they would not turn away any health-related coalition that requested services. The project staff was prepared to offer varied kinds of assistance, in response to the kinds of requests from coalitions (e.g., general capacity-building, information about evidence based practices, and so on). Therefore, the design took the form of a "dose-response" study, in which amounts of technical assistance utilized were correlated with levels of change in community effects.

DATA COLLECTION METHODS/MEASURES

Coalition Leaders: Annual Interviews

In 1996, semi-structured phone interview were conducted with the chairs or coordinators of each the 41 coalitions initially identified. The interview included questions about coalition operations and capacity, such as level of coalition activity, coalition structure, linkages developed between the local prevention coalition and other organizations, and so on. Nine respondents were the volunteer chairpersons of their coalitions. The remaining 32 respondents held a paid position on the coalition, 27 as coordinators. The following variables were used in this study:

Length of Existence as a Coalition. Information from coalition leaders and archival records were used to determine how long the coalition had been in existence.

Staff Hours. Leaders reported the average number of paid staff hours that the coalition had available on a weekly basis.

TA Interest. Coalition leaders were asked to rate how much technical assistance that they would want their coalition to receive in the coming year (1 = none, 4 = a lot) in nine different areas (e.g., recruiting new member organizations). This scale had an alpha of .83.

Linkages. Leaders reported on the extent of coalition contact during the past twelve months (on a scale from 0 = "none" to 3 = "weekly or more") with each of 12 different community constituencies (e.g., elementary schools, business community, police, and so on). A single linkage score was created for each coalition (Alpha = .84).

Capacity. Coalition capacity is a measure intended to gauge the organizational capacity of the coalition. The measure combines scores from three scales: Sectors, an 11 item scale that tallied the number of community sectors represented on the coalition, such as the business community, faith community; Formalization, a 12-item scale that tallied the number of structured elements in the coalition, such as written minutes or a plan of leader succession; Leaders, a count of the number of available leadership roles in the coalition. Capacity scale Alpha = .76.

Global Rating of Coalition Strength. Leaders rated the current strength of their coalition on a one item scale (1 = very weak; 4 = very strong).

See Florin et al. (2000) for previous use of several of the above measures.

Quarterly Coalition Leader Telephone Interviews

Coalition leaders also participated in brief telephone interviews focusing on the amount and perceived helpfulness of technical assistance services provided by the project. The quarterly interview was conducted over five consecutive quarters, covering the period of March 1997 through June 1998. Respondents were contacted the week following the end of a quarter.

Project Hours of Technical Assistance: For each coalition, we asked the coalition leader about his/her contact with each of the Project staff. We then calculated the total number of hours per quarter that the project provided to that coalition. We then calculated the mean hours per quarter across the project. If the respondent reported not using project services that quarter, we asked them to rate (1 = not at all, 2 = somewhat, 3 = very much) several potential reasons for not using services (e.g., not sure about the kind of services we need).

Quality of Project TA: For each Project staff member with whom they had received TA, we asked respondents to rate (1 = strongly disagree, 5 = strongly agree) the quality of the services (e.g., services were easy to access) they received from each project staff member during the previous quarter. The resulting three item scale had an alpha = .69. We then created a mean quality score for that quarter.

Non-Project Hours of Technical Assistance: We asked respondents to name up to five sources of technical assistance that they had accessed during the past quarter, and summed the hours of technical assistance. We calculated the mean hours per quarter across the project.

Quality of Non-Project TA: Respondents provided a single overall rating of their satisfaction with the non-project TA that quarter (1 = very dissatisfied, 4 = very satisfied).

We created mean scores for each of these variables, averaging across all the quarters for which we had data on a coalition. The average number of quarters in which we successfully collected data from coalitions was 3.7 quarters.

Key Informant Telephone Interview

Key informant interviews were initiated for coalitions that had participated in the initial Leader Interview. Although initiated in the spring of 1997, the questions covered the period starting with spring 1996. A short telephone interview was conducted with four key informants from the community (or service area) who rated the effects and impacts of the coalition in their community. Follow-up interviews were conducted in the fall of 1998.

Given the diversity of content areas of the coalitions we were studying, we designed a key informant sample to include one individual per community in each of these roles: (1) a school nurse; (2) a local health official, either a rural health center director or a hospital community relations representative; (3) a local government representative, typically the chair of the city or town council; and (4) the superintendent of schools. We expected that people in these roles ought to be knowledgeable about coalition activities in the areas of youth, health promotion/prevention, and governmental policies. Individuals in these roles should have been well-positioned to be aware of youth development and health promotion oriented initiatives in their communities.

Thirty-six coalitions of the original 41 coalitions were identified for administration of the initial Key Informant interview (i.e., five coalitions were either no longer active, had merged with another group, or indicated that they did not wish to continue with the evaluation project). Given our interest in 4 key informants for each coalition, this represented 144 interviews. We completed 141 of these interviews (one respondent refused to complete the interview, and two individuals could not be contacted after numerous attempts). Several coalitions served overlapping geographic areas, so that 31% of the 97 respondents rated more than one coalition. The respondents included 26 school nurses, 10 hospital community relations personnel, 12 rural health center directors, 25 chairs of town/city councils, and 24 superintendents of schools. Key informants had been in their positions for an average of 6 years and had worked in the area for an average of 15 years, ranging from less than 1 year to 55 years. Seventy-nine percent of the respondents lived in the area about which they were being questioned.

The specific variables examined were as follows:

Implementation Effects. This is a 7-item scale (alpha = .81) in which key informants rated the degree to which they "had personally noticed" any effects of the coalition (1 = none, 4 = a great deal) on several dimensions of community life that coalitions could be expected to influence (e.g., mobilizing youth and parents, community attitudes about alcohol and other drug problems, connections between agencies, policies of organizations, and resources available for prevention). If a Key Informant reported not being aware of the coalition, a "noticed no effects" was scored.

Collaboration/Coordination Among Coalitions. At time 2, questions about collaboration were added to the Key Informant interview. Respondents were asked "How much coordination or collaboration would you say is currently occurring among the different prevention and health promotion coalitions in your city, town, or area" (1 = none, 4 = a great deal). They also were asked to rate the level of coordination and collaboration that was happening a year ago. Key Informants who were unaware of a coalition's existence had their response coded as missing.

Final Sample Sizes

Given varied patterns of missing data across measures, coalition sample size differs depending upon the analyses performed: baseline description of leaders' ratings of their coalitions (n = 41), description of technical assistance utilization patterns (n = 38), pre-post differences in Key Informants ratings of coalitions (n = 34), pre-post differences in Key Informants ratings of coalitions as a function of amounts of technical assistance received, controlling for leader's baseline ratings (n = 31).

Copies of all measures are available from the first author.

RESULTS

At baseline, these coalitions had been in existence an average of 4.9 years, and were relatively active (an average of 11.05 coalition meetings in the past year). When asked, "how strong is your coalition?," 78% said "strong" or "very strong," while 22% said "weak" or "very weak."

Variables showing non-normal distributions were normalized through log transformations: length of time as a coalition, paid staff hours, project TA, non-project TA.

Utilization of Technical Assistance: Initial Contact

We examined the degree to which coalitions initially were engaged with the Project staff at their initial quarterly interview contact (n = 38). The coalitions were generally familiar with the project (37% very much; 57% somewhat, and 6% not at all). However, only 27% at that first contact reported utilizing project TA services.

Those coalitions that did not utilize the technical assistance were asked to rate the degree to which several factors contributed to their choice not to use the offered technical assistance services (1 = not at all, 2 = somewhat, 3 = very much). For 28.5% of the coalitions, the fact that they "had not decided what, if any, kinds of assistance that they needed" was "very much" a factor in deciding against using services, compared with "not in need of any services at this time" (23.8%), "received services from another source" (10%) and Project "does not have the kinds of services that we need" (10%). Thus, a failure to use services seems more associated with the lack of a clear sense of need, rather than skepticism about the offered services.

We correlated measures of coalition strength at time 1 with the leader's rating of the degree to which the coalition members "had not decided what, if any kinds of assistance that they needed." Lack of clarity about TA needs was negatively related to our composite time 1 measure of coalition capacity ($r = -.54$, $p < .001$), and negatively but not significantly with the leader's time 1 rating of the strength of the coalition ($r = -.36$, $p < .10$) or with key informants' time 1 ratings of coalition effects ($r = -.14$, $p < .53$). This suggests that coalitions with lower capacity are indeed more likely to struggle with defining needs, and less likely to utilize assistance.

Similarly, we correlated measures of coalition strength at time 1 with the leader rating of "not in need of any services at this time." There was no significant correlation between perceived lack of need for technical assistance and key informant ratings of coalition effects ($r = -.02$, $p < .90$), leaders' coalition strength ratings ($r = -.09$, $p < .65$), or coalition capacity ($r = .13$, $p < .54$). Coalitions reporting that they do not need TA services were not, by our other measures, more likely to be strong coalitions.

Patterns of Utilization of Technical Assistance: Project Life

In any one quarter, project staff provided services to approximately one-third of the coalitions that we contacted. Coalition leaders who used project services rated them very highly. For example, 93% of the respondents agreed "strongly" or "very strongly" that project staff were responsive to their needs. Coalition leaders also reported that 66% of these contacts were initiated by project staff. Across all quarters, however, 46% of coalitions failed to utilize any technical assistance services of project staff. The remaining coalitions received an average of 4.54 hours of assistance per quarter.

In contrast, approximately 79% of the coalitions reported receiving services from a non-project source, for an average of 15.1 hours per quarter. Coalitions utilizing non-project TA were also more likely to utilize project TA ($r = .38$, $p < .02$).

Predictors of Utilization of Technical Assistance

Overall, coalitions that were stronger at time 1 subsequently used more technical assistance. For example, time 1 capacity was significantly correlated with the amount of project TA subsequently utilized ($r = .37$, $p < .02$). The greater the capacity of the coalition at time 1, the greater extent to which the coalition used technical assistance services over the next 18 months. Other time 1 variables had negligible association with use of project technical assistance: Expressed interest in technical assistance ($r = .04$, $p < .81$), number of paid staff hours ($r = -.14$, $p < .42$), length of time in existence as a coalition ($r = .10$, $p = .57$), and linkage within the community ($r = .18$, $p < .30$).

For non-project TA, linkage within the community at time 1 was modestly correlated with utilizing more non-project TA ($r = .30$, $p < .08$). Again, there were no relationships between non-project TA utilized and expressed interest in technical assistance ($r = -.07$, $p < .71$), number of paid staff hours ($r = .23$, $p < .23$), and length of time in existence as a coalition ($r = .10$, $p = .57$). Coalition capacity was not significantly related to use of non-project TA ($r = .08$, $p < .65$).

Comparison of New versus Established Coalitions

New coalitions are often thought to be a particularly appropriate target for technical assistance. To examine this, we coded coalitions less than 2 years in existence as "new" coalitions and those more than 2 years in existence as "established" coalitions. New coalitions did not report significantly more overall interest in technical assistance but did differ in some specific self-identified needs. New coalitions were most concerned with "Building the knowledge and skills of individual members" ($t(37) = 2.174$, $p < .034$; new: $M = 3.36$, $SD = .50$; established: $M = 2.84$, $SD = .99$), while established coalitions were most concerned with "Planning for maintenance of the coalition over the long term" ($t(37) = -1.361$, $p < .182$; new: $M = 2.71$, $SD = .73$; established: $M = 3.12$, $SD = .97$). There was no difference between the groups in their utilization of technical assistance from either project or non-project sources.

Effects of TA on Coalition Collaboration and Community Effects

We examined changes in key informants' reports of coalition activity and effectiveness in their communities. Perceived coalition effects in their communities changed positively and this difference approached significance ($t(33) = 1.872$, $df = 1,34$, $p < .07$: baseline: $M = 1.85$, $SD = .57$; follow-up: $M = 1.97$, $SD = .62$). Perceived collaboration and coordination among coalitions increased significantly ($t(27) = 3.898$, $df = 1,27$, $p < .001$: baseline: $M = 2.66$, $SD = .77$; follow-up: $M = 3.05$, $SD = .77$).

However, amounts of technical assistance utilized were not significantly associated with levels of change in collaboration and community effects. To examine this, we dichotomized the TA variables, i.e., coalitions receiving no versus some project TA (n = 13, 18 respectively) and coalitions receiving minimal versus moderate non-project TA (n = 15, 16 respectively). Minimal is less than 7 hrs per quarter, X = 2.0; moderate is more than 7 hours, X = 19.9. Subsequent ANOVAs controlling for initial levels of collaboration and coalition effects showed no effects of project TA or non-project TA on time 2 levels of collaboration and coalition effects. (Full results are available from first author).

DISCUSSION

Provision of technical assistance to support prevention and health promotion interventions has become more commonplace, especially in the dissemination of empirically-based programs and in the promotion of community coalitions. This trend is likely to continue, as federal agencies pursue policies designed to turn "science into services." Enthusiasm for technical assistance has, however, quickly outstripped our knowledge of the optimal ways to design, promote, deliver and evaluate these interventions. Case studies of technical assistance to coalitions are few, and empirical works are even rarer. The lack of basic descriptive data about what actually happens when TA is delivered limits our ability to design more effective interventions and more rigorous studies of TA effects.

Ironically, many of the same questions now currently being raised about coalition initiatives would also apply to the use of technical assistance as an intervention. How should TA interventions be tailored to the individual circumstances of coalitions? What is the "penetration" of TA interventions, and do they reach the community-based organizations most in need of such help? What effect size is expected from a TA intervention and what is the minimal "dose" needed to obtain that effect? To what degree are TA interventions focusing on the "ecological" context of community coalitions? Unfortunately, the evidence base needed to address the above questions suffers from the same methodological challenges of coalition research: small sample sizes given communities as level of analysis, lack of longitudinal designs, difficulty in detecting small effects, and difficulty in tracking and measuring TA (Feinberg et al., 2002).

While this study represents only a modest first step, the results raise several important issues. First, coalitions with the least capacity were least likely to avail themselves of offered technical assistance. While ninety-four percent of coalitions were aware of the availability of project-supported technical assistance, only forty-six percent used it over the life of the project. Similarly, Feinberg et al. (2002) found some evidence that key community leaders were less likely to attend prevention training sessions if they came from communities with a lower level of community readiness for prevention. The two most frequent reasons given by respondents for low utilization were a lack of clarity about technical assistance needs and a lack of need for any TA "at this time." Technical assistance initiatives may have to work much more aggressively and systematically to reach coalitions that do not respond and might otherwise be overlooked. It takes preexisting capacity to focus on a mission, assess organizational and individual needs, and identify appropriate TA resources.

Providers of technical assistance quickly become aware that technical assistance need lies in the eyes of the beholder. Technical assistance providers must find a way to delicately balance a "consumer-oriented" approach where coalitions receive only what is asked for with a "diagnostic" approach where needs are judged against competence in performing essential "tasks." Florin, Mitchell and Stevenson (1993), for example, found that although members in 27 out of 35 coalitions rated their coalition low on "task focus," indicating dissatisfaction with the effectiveness of meetings, when asked about technical assistance needs, only 12% of members and 3% of leaders identified improving meeting effectiveness as a technical assistance need. It may in fact be easier for coalition members and leaders to acknowledge some needs rather than others. Data-based coalition "profiles" organized around dimensions of coalition functioning may be useful in broadening perspectives and tailoring technical assistance interventions to a coalition's specific circumstances (Dugan, 1996; Florin et al., 1993).

Second, technical assistance staff should think about the kind of effort they would need to expend to create the kind of effect they hope to achieve. In this study, there were no significant differences between coalitions that received "high" versus "low" amounts of TA intervention. However, there is little empirical evidence to estimate what appropriate "doses" might be. O'Donnell et al. (2000) reported spending from 2 to 4 hours of technical assistance per month over a nine month period in supporting Community-Based Organizations to replicate an evidence-based HIV intervention. However, efforts to assist coalitions with multi-component interventions may require much more time than replication of individual programs. TA systems should try to provide more descriptive data about their activities as a means of improving estimates of necessary levels of technical assistance. TA providers would have a much easier time assessing organizational or community readiness for technical assistance if they were clear about the minimal level of commitment needed from the organization or community to complete the recommended technical assistance process.

Third, TA systems need to specify the logic model of how particular TA interventions are to lead to particular outcomes. They need to especially take into account the "ecological context" in which they are intervening. The coalitions in this study reported receiving technical assistance from a variety of non-project sources, and rated these services highly. Providers of technical assistance must think about how their interventions mesh with other available resources. One promising area for coordination is in the provision of generic coalition capacity-building TA alongside the more problem-focused TA aimed at direct community action. This kind of coordination is challenging to achieve, since it is no less difficult to get TA providers to coordinate their efforts than it is to get coalitions to collaborate.

Fourth, although we did not find a relationship between amount of TA and improved coalition effectiveness over time, we did find that their effectiveness improved. Continued study is needed to examine qualitatively the coalition change process itself in order to better integrate TA with that natural, ecologically-grounded, process.

METHODOLOGICAL AND DESIGN CHALLENGES

Variability Across Coalitions in Their Programmatic Emphasis

The lack of significant effects of technical assistance could in part be due to using a single outcome measure across coalitions that were very diverse in their content focus (e.g., substance abuse, breast and cervical health, cardiovascular disease, and so on). Our logic model for the assessing the impact of TA rested on the following assumptions. Technical assistance would result in changes in the capacity of local coalitions, and greater effectiveness and/or reach of their prevention services. Key community leaders would notice increases in community coalitions with regard to their mobilization of the community, their ability to reach relevant constituencies, and their level of activity. Our questions to key informants asked generically about the effectiveness of the local coalition (e.g., mobilizing community resources, increasing connections between agencies, and so on). We tried to contact four key informants from diverse roles in each community in order to obtain a wide perspective on local coalitions. Given the small size of many of these communities, we presumed that coalition activity, if there were any, would be very visible to community leaders. However, without other data sources that could be compared with key informants' reports, we cannot be absolutely certain. As very diverse coalitions are encouraged to work across categorical problem boundaries, assessment of TA impact will become no less difficult and complex.

Coalition Evolution and Definitions of Success

Over the life of the project, a number of coalitions fell into inactive status according to our definition, i.e., had not met as a group within the last 12 months. However, there was great variability in how these coalitions became "inactive." One coalition lost funding, experienced a decline in member attendance, and faded out of existence. Another coalition that lost its major source of funding made a deliberate decision to cease to function as a separate entity and to merge with another group in its community. Do these contrasting endings have different effects on participants' willingness to get involved in future community development efforts? Is the decision of the latter coalition to merge an example of a coalition's failure to diversify its sources of support? Or is this a success story of how different groups in the same community can begin to coordinate their efforts? More time needs to be spent in tracking the natural life of coalitions, and understanding the implications of varied community roles over the long-term.

Sufficient Sample Size to Detect Effects

One problem with community interventions is the difficulty of collecting enough data to generate a sufficient effect. In this study, the sample size was adequate enough to detect a moderate pre-post change in the overall sample of coalitions. Is that too much of an effect size to expect? If the intervention could not produce a moderate-sized effect at the first step (i.e., changes in coalition collaboration and community mobilization), it seems unlikely that it would be powerful enough to ultimately change health outcomes several steps later.

Investigators in this area will have to be creative in their development of research strategies. Hansen and Collins (1994), for example, make several useful suggestions for improving power in research designs without increasing sample size. Another tack is the use of more standardized instruments and of a prevention research registry to make it easier to draw conclusions across projects (Brown, Berndt, Brinales, Zong, & Bhagwat, 2000). If TA systems tailor their interventions to the needs of specific constituencies, even larger sample sizes will be needed to detect the impact of more complex TA interventions.

Appropriate Time Frame for Examining Intervention Impacts

The logic model we have described above assumed that TA would change coalition capacity, which in turn would strengthen coalition actions such as mobilization and outreach. How long should this take? Developing the appropriate time frame requires some thought with regard to two questions: (a) how long should the program be in operation in order to have implemented sufficient enough activity to have caused an effect? and (b) is there a time lag between when sufficient programmatic activity occurred and when the change in outcome is expected to have occurred? One potential problem with evaluations of community-based interventions is that they may not allow sufficient time lag for the emergence of effects. As Rindskopf and Saxe (1998) indicate, this is especially true in assessing coalition or community partnership oriented initiatives: "It may take a year or more to organize all of the agencies interested in alcohol and other drug abuse to start planning how to coordinate their efforts, and several more years before they can actually coordinate their efforts" (p. 85). In the current study, the amount of time may well have been too short to have seen even intermediate effects. With externally funded coalition projects such as this one, however, the expectation is that results will be seen very quickly.

CONCLUSION

The movement towards technical assistance seems reminiscent of the early efforts at implementing community coalitions. The logic of the enterprise seems so compelling that it spurs on action even in the absence of empirical support. Perhaps some of the lessons that have emerged in years of work with community-based interventions are applicable here as well. One lesson has been to take a broad view of the ecology of an intervention, and the many factors that contribute to its implementation and success. Hopefully, technical assistance for community-based interventions can become more effective as we appreciate how the ecological context shapes, and is shaped by, efforts to support prevention interventions in the community.

REFERENCES

Brown, C. H., Berndt, D., Brinales, J. M., Zong, X., & Bhagwat, D. (2000). Evaluating the evidence of effectiveness for preventive interventions: Using a registry system to influence policy through science. *Addictive Behaviors, 25*, 6, 955-964.

Butterfoss, F. D., Goodman, R. M., & Wandersman, A. (1996). Community coalitions for prevention and health promotion: Factors predicting satisfaction, participation, and planning. *Health Education Quarterly, 23*, 65-79.

Center for Substance Abuse Prevention (n.d.). *CSAP's Prevention Pathways: Prevention Registry.* Retrieved July 10, 2002 from *http://preventionpathways.samhsa. gov/nrepp/default.htm*

Connell, J. P., Kubisch, A. C., Schorr, L. B., & Weiss, C. H. (Eds.). (1995). *New approaches to evaluating community initiatives: Concepts, methods, and contexts.* Washington, DC: The Aspen Institute.

Davis, D., Barrington, T., Phoenix, U., Gilliam, A., Collins, C., Cotton, D. et al. (2000). Evaluation and technical assistance for successful HIV program delivery. *AIDS Education and Prevention, 12*(5), 115-125.

Dishion, T. J., & Kavanagh, K. (2000). A multilevel approach to family-centered prevention in schools: Process and outcome. *Addictive Behaviors, 25*(6), 899-911.

Dugan, M. A. (1996). Participatory and empowerment evaluation: Lessons learned in training and technical assistance. In D. M. Fetterman, Kaftarian, S.J., & Wandersman, A. (Ed.), *Empowerment Evaluation: Knowledge and Tools for Self-Assessment and Accountability.* Thousand Oaks, CA: Sage.

Fawcett, S. B., Paine Andrews, A., Francisco, V. T., Schultz, J. A., Richter, K. P., Lewis, R. K. et al. (1995). Using Empowerment Theory in Collaborative Partnerships for Community Health and Development. *American Journal of Community Psychology, 23*(5), 677-697.

Feinberg, M. E., Greenberg, M. T., Osgood, D. W., Anderson, A., & Babinski, L. (2002). The effects of training community leaders in prevention science: Communities that care in Pennsylvania. *Evaluation and Program Planning, 25*(3), 245-259.

Florin, P., Mitchell, R., & Stevenson, J. (1993). Identifying training and technical assistance needs in coalitions: A developmental approach. *Health Education Research: Theory and Practice, 8*, 417-432.

Florin, P., Mitchell, R., & Stevenson, J. (2000). Predicting intermediate outcomes for prevention coalitions: A developmental perspective. *Evaluation and Program Planning, 23*, 157-163.

Francisco, V. T., Fawcett, S. B., Schultz, J. A., Berkowitz, B., Wolff, T. J., & Nagy, G. (2001). Using Internet-based resources to build community capacity: The community tool box [http://ctb.ukans.edu/]. *American Journal of Community Psychology, 29*(2), 293-300.

Gibbs, D., Napp, D., Jolly, D., Westover, B., & Uhl, G. (2002). Increasing evaluation capacity within community-based HIV prevention programs. *Evaluation and Program Planning, 25*(3), 261-269.

Green, L. W., & Kreuter, M. W. (2002). Fighting back or fighting themselves? Community coalitions against substance abuse and their use of best practices. *American Journal of Preventive Medicine, 23*(4), 303-306.

Hallfors, D., Cho, H., Livert, D., & Kadushin, C. (2002). Fighting back against substance abuse–Are community coalitions winning? *American Journal of Preventive Medicine, 23*(4), 237-245.

Hansen, W. B., & Collins, L. M. (1994). Seven ways to increase power without increasing N. In *NIDA Res Monogr* (Vol. 142, pp. 184-195).

Hays, C. E., Hays, S. P., DeVille, J. O., & Mulhall, P. F. (2000). Capacity for effectiveness: The relationship between coalition structure and community impact. *Evaluation and Program Planning, 23*(3), 373-379.

Kegler, M. C., Steckler, A., Malek, S. H., & McLeroy, K. (1998). A multiple case study of implementation in 10 local Project ASSIST coalitions in North Carolina. *Health Education Research, 13*(2), 225-238.

Kreuter, M. W., Lezin, N. A., & Young, L. A. (2000). Evaluating community-based collaborative mechanisms: Implications for practitioners. *Health Promotion Practice, 1*(1), 49-63.

Manger, T. H., Hawkins, J. D., Haggerty, K. P., & Catalano, R. F. (1992). Mobilizing communities to reduce risks for drug abuse: Lessons on using research to guide prevention practice. *Journal of Primary Prevention, 13*(11) 13-22.

McLeroy, K. R., Norton, B. L., Kegler, M. C., Burdine, J. N., & Sumayo, C. V. (2003). Community-based interventions. *American Journal of Public Health, 93*(4), 529-533.

Merzel, C., & D'Affitti, J. (2003). Reconsidering community-based health promotion: Promise, performance, and potential. *American Journal of Public Health, 93*(4), 557-554.

Minkler, M., & Wallerstein, N. (1997). Improving health through community organization and community building. In K. Glanz, F. M. Lewis & B. K. Rimer (Eds.), *Health behavior and education: Theory, research, and practice* (2nd ed., pp. 241-269). San Francisco: Jossey-Bass.

Mitchell, R. E., Florin, P., & Stevenson, J. F. (2002). Supporting community-based prevention and health promotion initiatives: Developing effective technical assistance systems. *Health Education & Behavior, 29*(5), 620-639.

O'Donnell, L., Scattergood, P., Adler, M., San Doval, A., Barker, M., Kelly, J. A. et al. (2000). The role of technical assistance in the replication of effective HIV interventions. *AIDS Education and Prevention, 12*(5), 99-111.

Rindskopf, D., & Saxe, L. (1998). Zero effects in substance abuse programs: Avoiding false positives and false negatives in the evaluation of community-based programs. *Evaluation Review, 22*(1), 78-94.

Roussos, S. T., & Fawcett, S. B. (2000). A review of collaborative partnerships as a strategy for improving community health. *Annual Review of Public Health, 21*, 369-402.

Saxe, L., Reber, E., Hallfors, D., Kadushin, C., Jones, D., & Rindskopf, D. (1997). Think globally, act locally: Assessing the impact of community-based substance abuse prevention. *Evaluation and Program Planning, 20*(3), 357-366.

Schinke, S., Brounstein, P. & Gardner, S. (2002). Science-Based Prevention Programs and Principles, 2002. DHHS Pub. No. (SMA) 03-3764. Rockville, MD: Center for Substance Abuse Prevention, Substance Abuse and Mental Health Services Administration.

Sorenson, G., Emmons, K., Hunt, M. K., & Johnston, D. (1998). Implications of the results of community intervention trials. *Annual Review of Public Health, 19,* 379-416.

Stevenson, J. F., Florin, P., Mills, D. S., & Andrade, M. (2002). Building evaluation capacity in human service organizations: A case study. *Evaluation and Program Planning, 25*(3), 233-243.

Wandersman, A., & Florin, P. (2003). Community interventions and effective prevention. *American Psychologist, 58*(6-7), 441-448.

Winett, R.A. (1995). A framework for health promotion and disease prevention programs. *American Psychologist, 50,* 341-350.

Aging and Prevention: New Approaches for Preventing Health and Mental Health Problems in Older Adults, edited by Sharon P. Simson, Laura Wilson, Jared Hermalin, PhD, and Robert E. Hess, PhD* (Vol. 3, No. 1, 1983). *"Highly recommended for professionals and laymen interested in modern viewpoints and techniques for avoiding many physical and mental health problems of the elderly. Written by highly qualified contributors with extensive experience in their respective fields." (The Clinical Gerontologist)*

Strategies for Needs Assessment in Prevention, edited by Alex Zautra, Kenneth Bachrach, and Robert E. Hess, PhD* (Vol. 2, No. 4, 1983). *"An excellent survey on applied techniques for doing needs assessments. . . . It should be on the shelf of anyone involved in prevention." (Journal of Pediatric Psychology)*

Innovations in Prevention, edited by Robert E. Hess, PhD, and Jared Hermalin, PhD* (Vol. 2, No. 3, 1983). *An exciting book that provides invaluable insights on effective prevention programs.*

Rx Television: Enhancing the Preventive Impact of TV, edited by Joyce Sprafkin, Carolyn Swift, PhD, and Robert E. Hess, PhD* (Vol. 2, No. 1/2, 1983). *"The successful interventions reported in this volume make interesting reading on two grounds. First, they show quite clearly how powerful television can be in molding children. Second, they illustrate how this power can be used for good ends." (Contemporary Psychology)*

Early Intervention Programs for Infants, edited by Howard A. Moss, MD, Robert E. Hess, PhD, and Carolyn Swift, PhD* (Vol. 1, No. 4, 1982). *"A useful resource book for those child psychiatrists, paediatricians, and psychologists interested in early intervention and prevention." (The Royal College of Psychiatrists)*

Helping People to Help Themselves: Self-Help and Prevention, edited by Leonard D. Borman, PhD, Leslie E. Borck, PhD, Robert E. Hess, PhD, and Frank L. Pasquale* (Vol. 1, No. 3, 1982). *"A timely volume . . . a mine of information for interested clinicians, and should stimulate those wishing to do systematic research in the self-help area." (The Journal of Nervous and Mental Disease)*

Evaluation and Prevention in Human Services, edited by Jared Hermalin, PhD, and Jonathan A. Morell, PhD* (Vol. 1, No. 1/2, 1982). *Features methods and problems related to the evaluation of prevention programs.*

Index

Acquired immunodeficiency virus
(AIDS), 70
Adaptation, ecological analogy, 5
Adolescent diversion project (ADP)
action condition, 35
advocacy activities, 45
attention placebo, 35
background, 32-35
bonding, 31
control condition, 35
correlation matrices, 41
court setting, 35
demographics, 39
early research, 32-35
ecological intervention, 29
effectiveness, 45
effects, 38-40
empirical path model, 40
family focus, 35,37,44
funding problem, 45
future research, 45
interactionist theory, 31
intervention models, 36
labeling measures, 37,39,44
learning theory, 31
method and results, 35-38
negative reputation, 43
participants, 36
path analytic model, 42
phase results, 33
program efficacy, 32-34
random assignment, 38
relationship condition, 35
replication study, 35-37
self-labeling, 37
services condition, 36
social control, 31
software program, 40

staffing models, 34
successful model, 34
theoretical background, 30
Adolescent Health Longitudinal
Survey, 62
Adolescents
community programs, 8
ecological interventions, 8
see also Youth Action Program
Air Force Academy, 52
Attention Scale, 61

Bandura, A., 6
Barker, R. G., 4
Behavior theory
environment, 6
setting, 4
Berks Behavior Rating Scale, 61

Camping, youth development, 56
Center for Substance Abuse
Prevention, 70
Children, Youth and Families At Risk, 17
Cognitive theory, social behavior, 6
Community
interventions, 2,16,70,84
psychology, 5,14
sustainability, 5
youth development, 51
Community Anti-Drug Coalition of
America, 70
Community-based organizations
(CBO), 70,81
Community coalitions
data collection, 73-77
funding, 70

BOOK ORDER FORM!

Order a copy of this book with this form or online at:
http://www.haworthpress.com/store/product.asp?sku=5170

Understanding Ecological Programming
Merging Theory, Research, and Practice

_____ in softbound at $19.95 ISBN: 0-7890-2459-4.
_____ in hardbound at $29.95 ISBN: 0-7890-2458-6.

COST OF BOOKS _____	❏ **BILL ME LATER:**
	Bill-me option is good on US/Canada/ Mexico orders only; not good to jobbers, wholesalers, or subscription agencies.
POSTAGE & HANDLING _____	
US: $4.00 for first book & $1.50 for each additional book	❏ **Signature** _____
Outside US: $5.00 for first book & $2.00 for each additional book.	❏ **Payment Enclosed: $** _____
SUBTOTAL _____	❏ **PLEASE CHARGE TO MY CREDIT CARD:**
In Canada: add 7% GST._____	❏ Visa ❏ MasterCard ❏ AmEx ❏ Discover
STATE TAX _____	❏ Diner's Club ❏ Eurocard ❏ JCB
CA, IL, IN, MN, NY, OH & SD residents please add appropriate local sales tax.	**Account #** _____
FINAL TOTAL _____	**Exp Date** _____
If paying in Canadian funds, convert using the current exchange rate. UNESCO coupons welcome.	**Signature** _____
	(Prices in US dollars and subject to change without notice.)

PLEASE PRINT ALL INFORMATION OR ATTACH YOUR BUSINESS CARD

Name		
Address		
City	State/Province	Zip/Postal Code
Country		
Tel	Fax	
E-Mail		

May we use your e-mail address for confirmations and other types of information? ❏Yes ❏No We appreciate receiving your e-mail address. Haworth would like to e-mail special discount offers to you, as a preferred customer. **We will never share, rent, or exchange your e-mail address.** We regard such actions as an invasion of your privacy.

Order From Your **Local Bookstore** or Directly From
The Haworth Press, Inc. 10 Alice Street, Binghamton, New York 13904-1580 • USA
Call Our toll-free number (1-800-429-6784) / Outside US/Canada: (607) 722-5857
Fax: 1-800-895-0582 / Outside US/Canada: (607) 771-0012
E-mail your order to us: orders@haworthpress.com

For orders outside US and Canada, you may wish to order through your local
sales representative, distributor, or bookseller.
For information, see http://haworthpress.com/distributors

(Discounts are available for individual orders in US and Canada only, not booksellers/distributors.)

Please photocopy this form for your personal use.
www.HaworthPress.com

BOF04